Bioethics and the Posthumanities

This interdisciplinary volume explores how posthumanist approaches can illuminate current issues in bioethics and considers the relevance of these issues for the humanities, including questions of autonomy and authorship, and notions of ethical and juridical responsibility in the context of a changing understanding of subjectivity.

With contributions from a variety of areas, including literature, philosophy, media, and policy-making, the book outlines the historical and philosophical development of posthumanism, and current key questions in bioethics. It generates a dialogue between bioethical approaches and the posthumanities, identifying ways in which posthumanist scholarship might be used to inform bioethical policy.

The book also looks more speculatively at the future, and the potential implications of technological developments which are only beginning to emerge. It uses posthumanism to look critically at the humanism underpinning de-extinction science, considers the ways in which technology is re-framing our social and political imaginaries, and asks about the identification of future posthumans.

Danielle Sands is Senior Lecturer in Comparative Literature and Culture at Royal Holloway, University of London.

Bioethics and the Posthumanities

Edited by Danielle Sands

Routledge
Taylor & Francis Group

LONDON AND NEW YORK

First published 2022
by Routledge
4 Park Square, Milton Park, Abingdon, Oxon OX14 4RN

and by Routledge
605 Third Avenue, New York, NY 10158

Routledge is an imprint of the Taylor & Francis Group, an informa business

British Library Cataloguing-in-Publication Data
A catalogue record for this book is available from the British Library

Library of Congress Cataloging-in-Publication Data
A catalog record for this book has been requested

ISBN: 978-0-367-89720-8 (hbk)
ISBN: 978-1-032-26367-0 (pbk)
ISBN: 978-1-003-02070-7 (ebk)

DOI: 10.4324/9781003020707

Typeset in Times New Roman
by Apex CoVantage, LLC

Contents

Contributors

Sarah Bezan is Postdoctoral Research Associate in Perceptions of Biodiversity Change at The University of York's Leverhulme Centre for Anthropocene Biodiversity in the UK. Her research focuses on the entangled social and ecological dimensions of species loss and revival in contemporary British, North American, and Australian literature and visual culture. She is currently at work on two book projects: *Dead Darwin: Necro-Ecologies in Neo-Victorian Culture* (under advance contract with Manchester University Press), along with a second monograph (in progress) that examines species revivalist representations of the woolly mammoth, great auk, dodo, Steller's sea cow, thylacine, and Pinta Island tortoise.

David Boden is a former PhD researcher at the University of Edinburgh.

Megen de Bruin-Molé (@Megen JM, she/her) is Lecturer in Digital Media Practice with the University of Southampton. Her book *Gothic Remixed* (Bloomsbury 2020) examines remix culture through the lens of monster studies, and her co-edited collection *Embodying Contagion* (UWP/Open Access 2021) explores how fantastical metaphors of contagion have infiltrated the way news media, policymakers, and the general public view the real world and the people within it. Megen is also an editor of the *Genealogy of the Posthuman*, an Open Access initiative curated by the Critical Posthumanism Network. Read more about her work on her blog: frankenfiction.com.

Ruth Chadwick is Professor Emerita at Cardiff University. From 2002 to 2013, she directed the ESRC Centre for Economic and Social Aspects of Genomics (Cesagen). From 2014 to 2018, she was a Professor of Bioethics at the University of Manchester. Ruth co-edits the journal *Bioethics* and has served on numerous bodies including the Council of the Human Genome Organisation. She is a Fellow of the Academy of Social

Sciences; of the Hastings Center, New York; of the Royal Society of Arts; of the Royal Society of Biology; and of the Learned Society of Wales. Her most recent publication is *This is Bioethics* (Wiley, 2021) co-authored with Udo Schüklenk.

Sarah Chan is Chancellor's Fellow in Ethics and Science Communicator in The Usher Institute at the University of Edinburgh. She has a wide interest in the ethics of medical research, including stem cells, embryo research, reproductive medicine, human enhancement, gene therapy, genetic modification, and animal ethics. She has published extensively in these areas. She was elected a Fellow of the Royal Society of Edinburgh Young Academy of Scotland in 2018.

Matt Hayler is Senior Lecturer in Contemporary Literature and Digital Cultures at the University of Birmingham and co-directs the Centre for Digital Cultures. His research explores human embodiment and our entanglement with technological artefacts and other non-human actors, drawing on the cognitive and digital humanities, 4E cognitive science, and posthumanist, post-phenomenological, and object-oriented philosophies (see, e.g. *Challenging the Phenomena of Technology*, 2015). He most recently worked on the AHRC-funded *Ambient Literature* project, and is currently looking at the ethics of human enhancement and the implications of posthumanism for real-world practices. Alongside Christine Daigle and Danielle Sands, Matt edits the Bloomsbury Academic series *Posthumanism in Practice*.

Stefan Herbrechter is *Privatdozent* (English and Cultural Theory) at the University of Heidelberg. He has published widely on English and comparative literature, cultural theory, and media studies. He is the author of *Posthumanism – A Critical Analysis* (2013) and *Before Humanity* (forthcoming in 2021), the Director of the *Critical Posthumanism Network* (http://criticalposthumanism.net/), and general editor of the *Critical Posthumanisms* series (https://brill.com/view/serial/CPH). For more information, see his homepage (stefanherbrechter.com).

Thomas Hobson is Research Associate at the Centre for the Study of Existential Risk, University of Cambridge, and Associate Member of the BioRISC Initiative at St Catherine's College, University of Cambridge. Tom's research is broadly concerned with understanding how scientists, policy makers, and societies imagine the future, the ways that existing technologies shape their visions of the future, and how they endeavour to secure a particular vision of the future through technology and innovation.

David Roden's research has addressed the relationship between deconstruction and analytic philosophy, philosophical naturalism, the metaphysics of sound, and posthumanism. His monograph *Posthuman Life: Philosophy at the Edge of the Human* (New York, 2014) explores the ethical and epistemological ramifications of Speculative Posthumanism. He also writes experimental fiction and concept horror works. His novella *Snuff Memories* is published by Schism Press (2021). He teaches at the Open University, UK.

Anna Roessing is a part-time PhD researcher at the University of Bath studying technology imaginaries in the Life Sciences. She is focusing on human enhancement technologies and is interested in understanding how expert and vanguard communities and policy actors imagine enhancement futures through existing and futuristic technology and scientific projects. Anna has a background in global health, biosecurity, and dual-use research governance in the life sciences, and is currently working as a Policy Manager in the NHS COVID-19 Vaccination Programme.

Danielle Sands is Senior Lecturer in Comparative Literature and Culture at Royal Holloway, University of London. She is the author of *Animal Writing: Storytelling, Selfhood and the Limits of Empathy* (Edinburgh, 2019) and the editor of *Philosophy and the Human Paradox: Essays on Reason, Truth, and Identity* (Routledge 2020). She received a British Academy Rising Star Engagement Award for her project "Posthumanities: Redefining Humanities for the Fourth Industrial Age," and is currently a Co-Investigator on the AHRC-funded network "The Philosophical Life of Plants." She co-edits the *Posthumanism in Practice* series for Bloomsbury Academic.

Michael Wee is a doctoral candidate in Philosophy at Durham University, where he is working on language, mind, and ethics in the later Wittgenstein. He is also an Associate Member of the Aquinas Institute at Blackfriars Hall, Oxford, and a Young Researcher Member of the Pontifical Academy for Life, the Holy See's advisory body on bioethics.

Introduction

Encounters Between Bioethics and the Posthumanities

Danielle Sands

If, as Francesca Ferrando provocatively contends, "[p]osthumanism is the philosophy of our time," sufficiently rich and radical to inform "an integrated redefinition of the human" (Ferrando 2019, 1), then it must speak beyond the humanities and outside the academy, engaging with those alienated by humanism and its legacies in order to transform legal, political, and ethical practice. However, it faces obstacles. Boasting varied "and even irreconcilable definitions" (Wolfe 2010, xi) alongside an astonishingly complex history, posthumanism risks appearing impenetrable and irrelevant, more interested in the minutiae of disciplinary debate than real-world impact. In the context of bioethics, a powerful, far-reaching discourse which has "left the clinic and entered the broader social world" (Zylinska 2009, 20), and which employs an entirely different language and methodology, the obstacles proliferate. Nonetheless, this volume makes the case that the posthumanities can and should inform the conversations around life, subjectivity, human nature, responsibility, and interspecies relations that underpin bioethical decision-making. It brings together posthumanism and bioethics for a series of exploratory, dialogical, and sometimes discordant encounters.

Origins and Aims of Posthumanism

Literary scholar Ihab Hassan coined the term posthumanism, then "a tendency struggling to become more than a trend" (Hassan 1977, 843), in his 1977 article "Prometheus as Performer: Toward a Posthumanist Culture?" Here, Hassan observed some of the ambiguities and ambivalences which would come to define posthumanism. For him, the posthuman names both a technologically driven modification of the human, in this case, the "cosmological extension of human consciousness" (844), and a departure from humanism, "not the literal end of man but the end of a particular image of us" (845), informed in part by Michel Foucault's reflections on the

DOI: 10.4324/9781003020707-1

contingency of "Man" at the close of *The Order of Things*. Subsequent decades have seen the popularisation of the term across a variety of disciplines and the development of two distinct branches of posthumanism, or rather, two very different (but occasionally intersecting) schools of thought which both make use of the term.

One branch, transhumanism, was first articulated by Julian Huxley in 1957, preceding Hassan's coinage of posthumanism by 20 years. Espoused by thinkers such as Hans Moravec (1988) and Nick Bostrom (2003), transhumanism (often confusingly labelled posthumanism in bioethical contexts) came to promote the use of technology to transcend the current limitations of the human via enhancement. This "techno-utopian" (Jansen, Leeuwenkamp, and Urricelqui 2021, 223) view, which increased in prominence during the 1980s, is effectively an "*intensification* of humanism" (Wolfe xv), combining the Enlightenment valorisation of scientific rationality with a commitment to technologies which promise liberation from perceived disorders and deficiencies, and even, perhaps, death. Humanities scholars have tended to query not the enhancement of the human in itself – which may well be liberatory or progressive in certain contexts, especially when separated from the teleological, humanist account of the human – but rather the triumphalist tone and widespread lack of political and philosophical reflectiveness in these projects. N. Katherine Hayles, for example, framed the problem as the "grafting of the posthuman onto a liberal humanist view of the self" (Hayles 1999, 286–87), where both discourses are underpinned by a disavowal of human embodiment. The stakes here are both philosophical – the misunderstanding of selfhood as solely immaterial – and political, as the prioritisation of narrowly defined conceptions of rationality has historically underpinned the marginalisation of those "othered" by race, sex, or species, through the association of these others with flesh, viscerality, or matter.

The second branch, critical posthumanism, is the primary focus of this volume, the multiple disciplines and media across which it operates comprising the posthumanities of the book's title. As a broad and varied "field of enquiry and experimentation" (Braidotti and Hlavajova 2018, 1), critical posthumanism emerges from the convergence "of posthumanism on the one hand and post-anthropocentrism on the other" (Braidotti 2019a, 32). While both transhumanism and critical posthumanism proceed from the acknowledgement of a certain condition – the "imbrication [of the human] in technical, medical, informatic, and exonomic, networks" (Wolfe 2010, xv) – they reach different conclusions. Transhumanism aspires to the posthuman – the transcendence of the physical limitations of the human and the creation of a new subject – whereas critical posthumanism "is only posthumanist, in the sense that it opposes the fantasies of disembodiment and autonomy" (Wolfe, xv). Resisting the temptations of recuperative apocalyp-ticism (Herbrechter 2013, 8) and premature triumphalism over humanism (Badmington 2003, 12), critical posthumanism builds upon the work of the primarily antihumanist "posts-" of the late twentieth and early twenty-first centuries. These discourses (including postmodernism, poststructuralism, postcolonialism, and postsecularism) exposed the processes by which Euro-pean modernity externalised its "colonial, violent, inhuman dimensions" (Jansen et al. 2021, 220) by projecting onto human others. Critical posthu-manism extends the acknowledgement of these "ghosts" (Herbrechter, 15) of humanism to the nonhuman – both animal and machine – exposing the "human" as "a historical 'effect'" (Herbrechter, 12) which "indexes access to privileges and entitlements" (Braidotti 2019a, 35).

Presupposing the indivisibility of the conceptual and the material, critical posthumanism looks to identify, resist, and disrupt mastery at all levels – an inevitably unfinishable task. It is particularly alert to the sta-bilisation, naturalisation, or appropriation of the figure of the "human," turning its focus to the contexts, networks, and entanglements from which it emerges. A constitutive supplementarity is rendered visible by posthu-manist logic: the "human" is not sovereign but co-constructed with and through its animal, human, and machinic others (see, for example, Ginn 2018, 413; Hayles, 2) and is thus primarily an embodied, relational being. For Cary Wolfe, this necessitates a double focus on the "thematics of the decentering of the human" and on "*how* thinking confronts that themat-ics" (Wolfe, xvi), in other words, the call to recast the processes and goals of thinking itself in order to meet this challenge. Critical posthumanism appeals to variety, both the amplification of the numerous human and non-human voices silenced by narrow notions of the "human," and a meth-odological diversity – theory, art, practice, and play – which counters the restricted account of knowledge underpinning traditional philosophical approaches.

Reimagining Bioethics

For those familiar with both bioethics and critical posthumanism, the lat-ter's relevance to bioethical questions and problems is clear. The four key principles of bioethics articulated by Tom L. Beauchamp and James F. Childress in their influential 1979 book *Principles of Biomedical Ethics* – autonomy, nonmaleficence, beneficence, and justice – continue to shape current formulations of bioethics. The single most important of these for contemporary bioethics, autonomy (Jennings 2009, 71), is challenged by critical posthumanism, which, in exposing the historical construction of the "human," problematises our assumptions about subjectivity (Braidotti 2019a), human nature (Herbrechter), nonhuman life (Ginn), disability (Taylor 2017), and what constitutes enhancement (Zylinska). Questioning a notion of personhood which falls short of articulating the networked human, critical posthumanism begins from the awareness that "bounded individualism . . . has finally become unavailable to think with" (Haraway 2016, 5), instead developing accounts of subjectivity which begin from bodies, their construction, and connections. In this way, critical posthumanism interrogates the foundations of our concepts of life and ethics. Regardless, it remains largely absent from bioethical conversations.

Traceable to both ancient and modern accounts of medical ethics – which presupposed the effectiveness of internal regulation – bioethics was formalised in the latter half of the twentieth century when the Nuremberg trials exposed the susceptibility of medical practitioners to the perpetuation of human suffering in the name of medical science (Obasogie and Darnovsky 2018, 4–5). It became clear that medical practice required an external regulatory authority and bioethics was born. In correcting for prevailing medical paternalism via an emphasis on autonomy and the centralisation of the "doctor-patient dyad" (Zylinska, 20), bioethics continues to be inherently individualistic, reinforcing individualism through a restricted and restrictive notion of "informed consent" (Obasogie and Darnovsky, 6). The force of this mode of bioethics, however, belies a more complex history. Guillaume Le Blanc, for example, identifies a conflicting type of bioethics, one not predicated on distinct agents but, as articulated by Van Rensselaer Potter in 1970, primarily ecological, and necessitating "a strong discussion about the interdependency of all living things" (Le Blanc 2015, 30). This tension has been lost by the reduction of bioethics to a mode of applied ethics which "brackets out environmental and animal ethics and tends to downplay socio-political, socio-economic and ecological inputs into human health" (Twine 2007, 511). Similarly, Eugene Thacker emphasises the plurality of bioethics, which combines, he argues, two strands – the philosophical and the technical – with the "often irresolvable questions" (2014, 180) of the former forced into action in a pragmatic, juridical framework. Bioethics presupposes, Thacker observes, that "the best way to respond to these conflicts is to develop broadly applicable rules that would hypothetically cover every instance" (180). It tends to follow the traditional models of philosophical ethics: deontology, virtue ethics, or, most widely, utilitarianism. These models are inescapably humanist, taking the autonomous human agent as the only unit of ethical value and regarding ethics as a practice of applying predetermined rules to particular cases. Such an approach, Margrit Shildrick suggests, is "out of touch with bodies themselves" (2005, 1–2), which posthumanist understandings of networked being make strikingly clear. Other critics examine the problems generated by bioethics' lack of reflexivity: subsumed by "neoliberal regulatory practices" (Apostolidou and Sturm 2016, 155), it is merely "a 'technological fix' to a technical problem" (Zylinska, 9). For Cary Wolfe, the reduction of ethics to "pragmatic expediency" (57) forecloses bioethics' ability to question its pre-theoretical assumptions; it becomes frictionless, "little more than the status quo with apologies" (60).

By inviting the posthumanities into bioethical conversations, this volume encourages a generative friction. It returns to the foundational questions which are frequently bypassed by bioethics – what is ethics? Which models of subjectivity are fit for purpose in the twenty-first century? How should we understand relations between human and nonhuman life? – and generates a dialogue between the posthumanities and the sciences to begin to answer them. Here, the figure of the posthuman is better understood as a productive "problem" than as a solution; it "consists of a series of interesting impasses that stall questions as they are currently formulated and require a new terrain" (Colebrook and Weinstein 2017, xxiv). This perception that philosophical thought should be disruptive and that such disruption serves a vital socio-political function, that is, that philosophy is "a means to *criticise* the present, to promote a reflective awareness of the present as being in crisis" (Critchley 2001, 73) marks out critical posthumanism as an inheritor of the so-called "continental" tradition in philosophy. In contrast to the "analytic" tradition, more naturally aligned with bioethics due to "analytic" thought's commitment to empiricism, critical posthumanism follows the continental belief "that the natural sciences do not provide human beings with their primary and most significant access to the world" (Critchley, 112). Building, therefore, on the supposition that philosophical scientism neglects the essential criticality of philosophy, critical posthumanism both denaturalises the beliefs underpinning bioethics (such as the philosophical and ethical primacy of the autonomous human subject) and challenges the presumption that "the question of the human" should be reserved for the sciences alone (Herbrechter, 169). A key contention of critical posthumanism, as chapters in this collection demonstrate, is that a normative, universalist morality rooted in personhood cannot elucidate, for example, the rapidly changing account of the human necessitated by genome editing, or by increased knowledge of the microbiome.

In response to these challenges, various thinkers in the humanities have begun to develop accounts of a "less enclosed" (Twine, 510) bioethics which does not proceed from liberal humanist ideals. Richard Twine outlines the possibility of a critical bioethics grounded in "interdisciplinarity, the avoidance of uncritical complicity with institutionalised relations of power, and self-reflexivity to subject matter" (Twine, 511). Shildrick (2019) draws on examples of chimerism and microchimerism to reveal the potential for biology to simultaneously disrupt the biosciences and the traditional philosophical subject. Informed by the philosophy of alterity and difference espoused by Emmanuel Levinas, Joanna Zylinska looks to replace the normative, universalist ethics of bioethics with a "nonsystemic" (Zylinska, xi) bioethics underpinned by recognition of, and responsibility to, difference, which resituates the subject materially, socially, and discursively. Such an approach demands a decoupling of ethics from "instant pragmatism" (178) and a radical interrogation of the notion of "life" unrestricted by assumptions of the uniqueness, superiority, and comprehensibility of human life. Thacker distinguishes between existing bioethics and "a parallel 'bioethics'" (176), a critical supplement to the former which reveals that the challenges faced by bioethics are not minor or procedural, but rooted in "more fundamental, philosophical questions" (178). Interrogating the Kantian bedrock of bioethics, Thacker, like Zylinska, develops a non-programmatic account of bio-ethics which reflects upon the "bio" in light of new technologies and which regards bodies relationally rather than as discrete objects to be treated. This "bio-ethics," which resists a prescriptive or teleological account of the human, looks to rework the discipline of more traditional bioethics from the bottom-up, shifting focus from law to practice to determine, discursively and experimentally, "what counts as a 'body'" (Thacker, 191). This volume follows the claim made by both Zylinska and Thacker that ethics need not be understood as an application of pre-existing rules, but may also be "a form of critique" (Thacker, 191). Like Thacker, it imagines a "counterpoint" (191) between the prevailing model of bioethics and another, one simultaneously more critical and more creative. Departing from Thacker, however, it turns explicitly to the posthumanities in order to find this new perspective.

A comparable transition is also underway in scholarship on health. The development of disciplines such as the biohumanities, itself an explicitly "critical enterprise" (Stotz and Griffiths 2008, 40), and transdisciplinary approaches to health such as the One Health initiative, which proceeds from the interconnection between humans, animals, and their environments (Rock 2017), indicates a turn towards a more reflexive, and potentially less anthropocentric, approach to human health. David Chandler contends that the COVID-19 pandemic has exposed the limitations of the resilience model, giving force to "an underlying shift away from the values of liberal humanism" (Chandler, n.p.). This shift is visible in the adoption of models of distributed agency which do not begin from the human subject (see, for example, Mills on HIV, 2017; Garnett on toxicity, 2017; Lynch and Cohn on blood, 2017); in a model of health and harm reduction which "strives to attune to . . . human and nonhuman *processes*, to reconfigure bodies in 'healthier' ways" (Dennis 2017, 337); and in challenges to the instrumentalisation of nonhuman life in health and medicine (Svendson 2017). Such accounts look to redefine health relationally, sometimes via explicitly posthumanist accounts, rather than as a property of a singular subject. In a 2019 article Gavin J. Duff and Cameron Andrews trace the "slow burn" (Duff and Andrews 2019, 130) of the posthumanist turn in health studies, illustrated, they describe, by a focus on material assemblages; the perception of bodies as vital and performed; the redefinition of health as affective experience; and experimentation with alternative vocabularies for bodies, health, and illness. Posthumanism, they claim, offers a "new wonderment" (130), in addition to informing political strategies which might facilitate "greater social unity and equality" (127).

These explorations of the transformative potential of posthumanism are exciting, but should be accompanied by scepticism towards the claims that humanism might be definitively overcome or that posthumanism can be distilled into a set of principles to be followed. Rather, taking posthumanism seriously entails an ongoing "working-through of humanist discourse" (Badmington, 22), where the "post" is not a simple temporal break but "implies a continuity, a discontinuity, and a transcendence (in the literal meaning of exceeding)" (Ferrando, 66). In practical terms, this means that a bioethics informed or inflected by posthumanism must proceed doubly, operating within existing discourses to inform current decision-making *and* interrogating the very formulation of these issues, exploring multiple, as yet inexpressible, futures. It should resist the unreflexive reproduction of a bioethics which aligns perfectly with prevailing public values, instead questioning those very values and identifying both what is missed by them and what is reproduced in error or to the detriment of humans and other beings.

Chapter Outlines

The essays in this volume navigate biology, philosophy, literary studies, political theory, and cultural studies, building and operating "tool-kits" (Braidotti 2019a, 48) from the posthumanities to engage with bioethical questions. The volume as a whole does not appeal to consistency of approach or perspective, but rather to the value of multiple, sometimes dissonant, voices.

The volume is divided into three parts. Part I addresses current issues in bioethics. Beginning with the prevailing notion of the posthuman in bioethics – which takes a technologically enhanced human as its goal – Michael Wee's chapter examines the difference between therapy and enhancement, demonstrating that the distinction is not an absolute one. Stressing the interplay between society and biology, Wee turns to the social model of disability in order to develop a contextual approach, based on social consideration of equity, to determine the value of different kinds of enhancement. Tracking its emergence and development, David Boden and Sarah Chan critique the bioethical account of the posthuman which, they argue, both re-centres the human and reinforces hegemonic norms about human embodiment and "species typical" flourishing. Instead, adopting a critical posthumanist lens and taking human heritable genome editing as an example, Boden and Chan look to a posthumanism which challenges exclusionary definitions of the "human" and generates reflection on the conception of a "good life." Focusing on the history of ethics in the Human Genome Organisation, Ruth Chadwick investigates the development of bioethics and its possible futures. Examining key changes in the history of "gen-ethics" – including the "communitarian turn," advances in epigenomics, and advancing knowledge of the similarities between the human genome and the genomes of other species – Chadwick demonstrates how the broadening of "gen-ethics" (and, by association, bioethics more generally) illustrates its increasing openness to the posthumanities.

Part II places bioethics and posthumanism in dialogue. Using examples including the controversy surrounding "CRISPR babies," Anna Roessing and Thomas Hobson identify a lack of sustained engagement with the political dimensions of human enhancement technologies. Their response, drawing on Science and Technology Studies as framed within the posthumanities, considers the ways in which cultural and political values, foreign policy, and geopolitics direct technological developments, demonstrating how theoretical tools from the posthumanities can inform a bioethics which is able to reflect on the construction of technological imaginaries. In his chapter, Stefan Herbrechter identifies the coincidence between a shift in life sciences towards a new microbiology and the turn towards the nonhuman in the humanities, which together ask with renewed urgency: What is technology? What is the human? What is life? Revealing the limitations of transhumanism, Herbrechter argues that a posthumanist approach, one which takes the critique of humanism seriously and looks to a post-anthropocentric perspective, is necessary to respond to the technological and eco-biological challenges that we face. In her chapter, Megen de Bruin-Molé explores the relationship between the theorisation of human bodies and of nonhuman creations (including media and artworks), revealing the parallels between our perceptions of bodily and intellectual autonomy. Addressing the problematisation of notions of authorship and ownership through the "open access" and "remix" movements, understood as instantiating a posthumanist approach, de Bruin-Molé argues that textual ethics is a useful tool through which to understand and reframe bioethics.

Part III looks more speculatively towards the future, and the potential implications of technological developments which are only beginning to emerge. Sarah Bezan's chapter presents a posthumanist critique of de-extinction science by exposing the ways in which anthropocentrism – in the form of a longing for macroevolutionary authority – underpins biotechnological initiatives which look to re-vivify extinct species. Exposing entrenched humanism on both sides of the debate, Bezan argues that attitudes to de-extinction reflect our perceptions of nonhuman life more broadly, and that contemporary art and media create the cultural space to negotiate the issues raised by de-extinction and to conceive of futures less bounded by anthropocentrism. In his chapter, Matt Hayler contends that posthumanist accounts of the self – which substitute notions of sovereignty and autonomy for an increased awareness of intra- and inter-subjective entanglements – can be used to inform accounts of moral responsibility which respond to current bioethical questions. Focusing on the challenges raised by an increasing understanding of the microbiome and its impact on human behaviour, Hayler argues that posthumanism cannot avoid a rethinking of prevailing conceptions of culpability, and can offer a basis to generate different formulations of punishment and justice. In the final chapter, David Roden argues that Speculative Posthumanism – which challenges human-centric thinking about the future implications of technology – reveals the parochialism of bioethics. Contending that the imposition of anthropocentric filters which disregard non-anthropoform posthumans cannot be justified when applied to deep time, he argues that a bioethics of posthumanity is not possible until we are in a position to encounter, observe, or become posthumans.

Bibliography

Andrews, Gavin J., and Cameron Duff. 2019. "Matter Beginning to Matter: On Posthumanist Understandings of the Vital Emergence of Health." *Social Science and Medicine* 226: 123–34.

Apostolidou, Sofia, and Jules Sturm. 2016. "Weighing Posthumanism: Fatness and Contested Humanity." *Social Inclusion* 4(4): 150–59.

Badmington, Neil. 2003. "Theorizing Posthumanism." *Cultural Critique* 53: 10–27.

Beauchamp, Tom L., and James F. Childress. 1979. *Principles of Biomedical Ethics.* New York and Oxford: Oxford University Press.

Bostrom, Nick. 2003. "A History of Transhumanist Thought." *Journal of Evolution and Technology* 14(1): 1–25.

Braidotti, Rosi. 2013. *The Posthuman.* Cambridge and Malden, MA: Polity Press.

———. 2019a. "A Theoretical Framework for the Critical Posthumanities." *Theory, Culture & Society* 36(6): 31–61.

———. 2019b. *Posthuman Knowledge.* Cambridge and Medford, MA: Polity Press.

Braidotti, Rosi, and Maria Hlavajova. 2018. "Introduction." In *Posthuman Glossary,* edited by Rosi Braidotti and Maria Hlavajova, 1–14. London and New York: Bloomsbury Academic.

Chandler, David. 2020. "Coronavirus and the End of Resilience." In *E-International Relations.* Coronavirus and the End of Resilience (e-ir.info).

Colebrook, Claire, and Jami Weinstein. 2017. "Preface: Postscript on the Posthuman." In *Posthumous Life: Theorizing Beyond the Posthuman*, edited by Claire Colebrook and Jami Weinstein. New York: Columbia University Press.

Critchley, Simon. 2001. *Continental Philosophy: A Very Short Introduction*. Oxford: Oxford University Press.

Dennis, Fay. 2017. "The Injecting 'Event': Harm Reduction Beyond the Human." *Critical Public Health* 27(3): 337–49.

Ferrando, Francesca. 2019. *Philosophical Posthumanism*. London and New York: Bloomsbury Academic.

Garnett, Emma. 2017. "Enacting Toxicity: Epidemiology and the Study of Air Pollution for Public Health." *Critical Public Health* 27(3): 325–36.

Ginn, Franklin. 2018. "Posthumanism." In *The Edinburgh Companion to Animal Studies*, edited by Lynn Turner, Undine Sellbach, and Ron Broglio, 413–29. Edinburgh: Edinburgh University Press.

Hassan, Ihab. 1977. "Prometheus as Performer: Toward a Posthumanist Culture?" *The Georgia Review* 31(4): 830–50.

Haraway, Donna. 2016. *Staying with the Trouble: Making Kin in the Chthulucene*. Durham, NC: Duke University Press.

Hayles, N. Katherine. 1999. *How We Became Posthuman: Virtual Bodies in Cybernetics, Literature and Informatics*. Chicago and London: University of Chicago Press.

Herbrechter, Stefan. 2013. *Posthumanism: A Critical Analysis*. London and New York: Bloomsbury Academic.

Jansen, Yolande, Jasmijn Leeuwenkamp, and Leire Urricelqui. 2021. "Posthumanism and the 'Posterizing Impulse.'" In *Post-Everything: An Intellectual History of Post-Concepts*, edited by Herman Paul and Adriaan van Veldhuizen, 215–33. Manchester: Manchester University Press.

Jennings, Bruce. 2009. "Autonomy." In *The Oxford Handbook of Bioethics*, edited by Bonnie Steinbock. Oxford: Oxford Handbook of Bioethics – Oxford Handbooks.

Le Blanc, Guillaume. 2015. "A Brief History of Bioethics." In *The Care of Life: Transdisciplinary Perspectives in Bioethics and Biopolitics*, edited by Miguel de Beistegui, Giuseppe Bianco, and Marjorie Gracieuse, 25–32. London and New York: Rowman and Littlefield.

Lynch, Rebecca and Simon Cohn. 2017. "Beyond the Person: The Construction and Transformation of Blood as a Resource." *Critical Public Health* 27(3): 362–372.

Mills, Elizabeth. 2017. "Biological Precarity in the Permeable Body: The Social Lives of People, Viruses, and Their Medicines." *Critical Public Health* 27(3): 350–61.

Moravec, Hans. 1988. *Mind Children: The Future of Robot and Human Intelligence*. Cambridge: Harvard University Press.

Obasogie, Osagie K., and Marcy Darnovsky. 2018. "Introduction." In *Beyond Bioethics: Toward a New Biopolitics*, edited by Osagie K. Obasogie and Marcy Darnovsky, 1–14. Oakland, CA: University of California Press.

Rock, Melanie J. 2017. "Who or What is 'the Public' in Critical Public Health? Reflections on Posthumanism and Anthropological Engagements with One Health." *Critical Public Health* 27(3): 314–24.

Shildrick, Margrit. 2005. "Beyond the Body of Bioethics: Challenging the Conventions." In *Ethics of the Body: Postconventional Challenges*, edited by Margrit Shildrick and Roxanne Mykitiuk, 1–26. Cambridge, MA and London: MIT Press.

———. 2019. "(Micro)chimerism, Immunity and Temporality: Rethinking the Ecology of Life and Death." *Australian Feminist Studies* 34(99): 10–24.

Stotz, Karola, and Paul E. Griffiths. 2008. "Biohumanities: Rethinking the Relationship between Biosciences, Philosophy and History of Science, and Society." *The Quarterly Review of Biology* 83(1): 37–45.

Svendson, Mette N. 2017. "Pigs in Public Health." *Critical Public Health* 27(3): 384–90.

Taylor, Sunaura. 2017. *Beasts of Burden: Animal and Disability Liberation.* New York and London: The New Press.

Thacker, Eugene. 2014. *Biomedia.* Minneapolis, MN and London: University of Minnesota Press.

Twine, Richard. 2007. "Thinking Across Species – A Critical Bioethics Approach to Enhancement." *Theoretical Medicine and Bioethics* 28: 509–23.

Wolfe, Cary. 2010. *What Is Posthumanism?* Minneapolis and London: University of Minnesota Press.

Zylinska, Joanna. 2009. *Bioethics in the Age of New Media.* Cambridge, MA, and London: MIT Press.

Part I

Bioethical Challenges

1 Therapy, Enhancement, and the Social Model of Disability

Michael Wee

My subject is the ethics of enhancement, which arguably is one of the foundational issues in transhumanist and posthumanist bioethics. It is worth saying at the outset that while there are no fixed borders between transhumanism and posthumanism, one useful way of picturing the relationship between the two is that of process and endpoint. According to a common definition, attributed to Max More, "Transhumanism is a class of philosophies that seek to guide us towards a posthuman condition (Aydin 2017, 307)." This paper is therefore, in part, a critique of potential teething issues in the prospective journey to posthumanism – understood in an aspirational sense rather than as a critical tool for engaging with ethical and environmental issues, as the term is used elsewhere in this volume. The posthuman aspirations I have in mind are those that generally take technologically induced enhancement to be an important means of transcending the limits of the human as we know it, so that we might "evolve" into radically new kinds of beings (Aydin 2017).

In this paper, I hope to demonstrate that the very notion of enhancement is, in fact, far from straightforward by suggesting two perspectives for approaching the subject – the therapy-enhancement distinction and the social model of disability. On the surface, the two conceptual models appear to be pushing in opposite directions from each other. The former seems to rely on some objective view of human nature or "normal functioning," which can thereby function as a yardstick for characterising interventions as either therapy or enhancement. For that reason, the distinction has been particularly important in discussions about gene-editing. By contrast, the social model of disability is inherently suspicious of any notion of "normal functioning." Disability is seen primarily or exclusively as the product of social barriers, rather than any physical or mental impairments. The model has been instrumental in driving social change for greater inclusion of those with disabilities (Anastasiou and Kauffman 2013, 441–45).

Although both ideas begin in different places, they shall converge somewhere in the argument I will develop, which will consider each in

DOI: 10.4324/9781003020707-3

turn in order to illustrate certain difficulties with the ethical evaluation of enhancement. While therapy and disability might, in their own ways, seem to be the opposites of enhancement, the reality is that enhancement is a somewhat slippery notion and that, in turn, problematises the "post" in "posthuman."

Therapy and Enhancement

What is the therapy-enhancement distinction and is it a useful ethical tool? I would like to suggest that it is, but only if we are clear about the context in which the distinction is meant to operate. The distinction, I shall argue, is best thought of not as an absolute model for classifying interventions, but as a relative criterion for certain policy aims.

To understand the distinction, it helps to clarify its different levels. One might seek to distinguish between therapy and enhancement *conceptually*, and therefore one would need an account of disease and, perhaps, one of normal human functioning as well. Then there is the question of what *practical purpose* the distinction serves. Simply distinguishing conceptually between therapy and enhancement does not of itself produce any argument for the immorality of enhancement. However, there might be practical implications, such as whether enhancement interventions should be part of publicly funded healthcare or insurance coverage. In what follows, I shall focus on the place of enhancement in healthcare.

Many, I think, would agree that the therapy-enhancement distinction has some intuitive appeal for explaining the aims of medicine: What else is medicine for if not fighting disease, and restoring the body to its full integrity? That is something of a paradigmatic picture (Eijk 2017). However, medicine is, arguably, much more than that. Palliative medicine, for example, does not necessarily seek to fight disease, and is called upon precisely when one cannot cure someone or restore her to full integrity. Instead, it is aimed at improving and maintaining quality of life via technical interventions, and is often complemented by the work of professionals from other disciplines besides medicine. This is indicative of the fact that, even if we took normal bodily functioning to be a starting point, as the practice of medicine develops it inescapably serves more complex social ends as well. In the case of palliative medicine, being able to die well is one such social end.

More germane to our topic is the field of preventive medicine. Diseases, after all, generally have a flipside – good health. So long as one accepts the legitimacy of preventive medicine as part of healthcare provision, then it is difficult to deny that enhancement has a role to play in medicine. Vaccines are perhaps the prime example of an intervention that is widely accepted as a legitimate part of medicine, and that does not aim to restore health to full integrity or to a normal species-level of functioning. In fact, it is the recognition of a species-level inadequacy in our immune system. It seeks to enhance, not heal.

In a similar vein, Austriaco (2017, 46–47) points out that statins are commonly prescribed to lower cholesterol levels and decrease the risk of cardiovascular disease, and that many clinicians believe that low-density lipoprotein (LDL) cholesterol levels should be lowered far below the normal human range – to 70 or 50 mg/DL (the US adult average is 119 mg/dL) – to be effective as a form of prevention. If this is not unethical, then the upshot is that enhancement is not only morally permissible but also an integral part of healthcare. Therefore, at the level of practice, the therapy-enhancement distinction does not help delineate medical from non-medical practices. However, would it be hasty to write it off completely?

Normal Functioning and the Social Ends of Medicine

One possible response to the line of argument developed earlier is to say that enhancements of the sort we have mentioned – be they vaccines or statins or increasing resilience – are still therapeutic in their aim, insofar as some disease or pathology still forms part of their *raison d'être*. The distinction that really matters is between preventive and non-preventive enhancements (Juengst 1997). What is not envisaged as part of medicine is enhancement that bears no relation whatsoever to any kind of pathology. So an enhancement to prevent cancer might be evaluated differently from an enhancement that sought to remedy short stature, or to bestow hearing abilities that are "off the charts." The dispute over what exactly is a disease may be difficult to settle conclusively, but surely there are paradigm cases of disease, and paradigm cases of conditions which are not pathological. Might there not be something to the idea that one can distinguish between enhancements that are preventive – or "pre-emptively therapeutic" – and those that are altogether non-preventive?

While it is beyond the scope of this paper to enter into discussion about the nature of disease, such a response suggests certain points for further consideration. Firstly, it is simply difficult to deny, in practice, that quality of life is a legitimate aim of medicine. Although at the level of an individual instance of an intervention, preventive medicine is disease-oriented, as a wider practice there have long been questions concerning the cost-effectiveness of preventive medicine. It is not always the most efficient use of resources for combating disease, as treating disease in individuals post-diagnosis could well be cheaper on the whole than a large-scale prevention programme. However, if medicine is not simply about targeting disease but also about improving positive health, then even costlier forms of preventive medicine could be justified (Cohen and Neumann 2009). Preventive medicine therefore brings with it a paradigm shift in medicine, as does palliative medicine which is also concerned with quality of life. So even if one could distinguish between preventive and non-preventive enhancements, this may not matter so much as long as a particular enhancement conceivably improves positive health.

Another consideration is that it can be difficult to determine the precise aim of an enhancement. Enhancements such as increasing one's lifespan or physical strength beyond current human norms may not, on the face of things, seem to be disease-oriented in the way that vaccines or statins might be. This is partly due to the indeterminacy of many proposed and largely speculative posthumanist enhancements, a problem that I will address in greater detail later. But just what is meant by increasing one's lifespan – by what means shall this be achieved? While certain characteristics – lifespan, memory, intelligence – may, in abstract, seem attractive as candidates for enhancement, they are unlikely to be easily delineated in biology and are not susceptible to "simple cause and effect predictability" – especially given how interrelated bodily and psychological characteristics are (Fitzgerald 2008, 42). If that is the case, it strikes me as unlikely that any intervention to increase one's lifespan or physical strength would not, on some level, prevent the onset of disease. So even if one wished to apply a strict prevention criterion for enhancements, that would not necessarily exclude many potentially contentious enhancements.

Ultimately, it seems that there is no intrinsic moral problem with enhancement, and furthermore, enhancement is quite consistent with fairly uncontroversial areas of medical practice. To restrict medicine purely to fighting disease and the restoration of bodily integrity would come at a cost that neither philosophers nor policymakers are likely to be willing to pay.

Yet, as I mentioned earlier, how valuable the therapy-enhancement distinction is depends on what work one would like it to do. As an absolute criterion for determining the boundaries of medicine, it would be unrealistic. As a conceptual classification, there are difficulties relating to the definition of disease. However, I would suggest that it is best thought of as a rudimentary point of departure, from or around which more nuanced standards emerge about the aims and boundaries of medicine. Medicine might be seen, not as a unified practice, but as a family of interrelated practices. Perhaps its core purpose is therapy – the reversal of ill health, which was our paradigmatic picture – but around that core are practices aiming at alleviating symptoms even without the prospect of cure, or maintaining positive health in body and mind, or even making people "healthier than healthy," to borrow a phrase (Austriaco 2017, 49).

In this way, the notion of the normal range of human functioning might still have a role to play in ethics and healthcare policy. To use normal functioning as a yardstick for determining the therapeutic core of medicine is not necessarily to be committed to acknowledging some special metaphysical importance about the natural. But as Daniels (2000, 318–19) puts it, "the natural baseline has become a focal point for convergence in our public conception of what we owe to each other by way of medical assistance or healthcare protection." It is about developing "fair terms of cooperation." In a word, the question at hand is one of equity.

The usefulness of the therapy-enhancement distinction is therefore something in between rhetoric and absolute criterion. It is less helpful in determining what is absolutely excluded, but more helpful in adjudicating what comes first as a matter of priority. In very simplistic terms, interventions that are most directly therapeutic are more likely to have greater priority, with those that are preventive ranking lower. Still more elaborate enhancements that seem disproportionate to their preventive effects – say, a speculative enhancement that not only prevented age-related muscle atrophy but also bestowed super strength – might be even lower in priority, even if not absolutely incompatible with the ethos of medicine.

Such a model preserves the instinct that therapy is at the core of medicine, while recognising the reality that enhancement is not of a wholly different character. Because equity is acknowledged more explicitly as driving the distinction, as opposed to biological function and statistical norms, it can also accommodate the perspective that the aims of medicine are at least partially socially determined, and this may be particularly pertinent to borderline cases of disease.

Consider, for example, how there are numerous possible genetic mutations which many of us carry without realising it, simply because they are harmless. However, there are also the more severe and debilitating ones like Tay-Sachs disease. Where do the mutations that cause colour vision deficiency (commonly referred to as "colour blindness") stand? The condition does not pose a huge difficulty in many life circumstances, though depending on its severity one may not become a pilot, for instance. But imagine a different set of social circumstances where colour vision deficiency makes it impossible for someone to hold fairly ordinary jobs – a hypothetical society which relied heavily on colours for communication and coordination. Or perhaps it is a futuristic world where piloting is just as commonly needed as driving is in ours. If we understood the therapy-enhancement distinction through a model of equity, rather than an exclusively biologistic view of disease, then we would give more weight to social ends in medicine in policymaking, while recognising that these ends can vary with changing social conditions.

Posthuman Disability

I have argued that enhancement is not easily separable from therapeutic aims in medicine. However, we can try to adjudicate its place in relation to more paradigmatic forms of therapy by taking into account the social ends of medicine, such as equity. This would rely on a more modest therapy-enhancement distinction. As we have seen, this also promotes sensitivity to socially constructed disability, which provides us with a segue into a slightly different discussion. If enhancement is not intrinsically wrong, how else might we assess its other ethical dimensions? In the remainder of this book, I would like to suggest how the social model of disability might offer some valuable perspectives on the subject.

The social model of disability, which is both a concept and a political tool, captures a very simple, if powerful idea – disability is constructed by forms of social organisation that exclude certain groups of people by the barriers or structures it erects, whether they are physical (like steps) or intangible (such as employment opportunities and expectations). It is these structures that cause disability, not any form of physical or mental impairment (Anastasiou and Kauffman 2013).

What might we say about enhancement with this model of disability in mind? Just as the distinction between therapy and enhancement is not straightforward, in my view the social model of disability suggests that the difference between enhancement and disability is not clear-cut either. Unless an equitable distribution of enhancement-directed resources is achieved, might the comparatively small number of enhanced individuals end up being disabled by social structures?

In the contemporary developed world, one is familiar enough these days with a situation where an economy does not quite have enough suitable jobs for scores of overqualified university graduates. With enhancement, things are potentially more dramatic than that, as the following considerations might suggest. If, for example, a small number of people were able to extend their lifespan by 50 years, would they find adequate family and social structures able to support and accompany them in their extended longevity, given that as things are people can find themselves spread thinly between parents and grandparents, to say nothing of great-grandparents? The impact of multiple "super-supercentenarians" on family or community dynamics is difficult to foresee. Would someone with greatly enhanced cognitive abilities be able to function well in a regular work environment, for instance, or would they instead be disabled by the speed at which things progressed relative to their own mental powers? Would someone with hearing abilities going well beyond the normal human range be able to cope with being in a crowded place, or be able to appreciate an orchestra, or be at risk of being overwhelmed by the sheer volume of sounds and, indeed, in the latter case, mistakes that no human ear can hear at present? Or think of another cultural artefact such as a van Gogh artwork – step close enough and one can see the thick, furious brushstrokes for which the painter is well known, but step back a little and one sees the overall effects of those brushstrokes in the context of the whole picture. Perhaps someone whose vision had been enhanced so greatly that they could perceive details more finely than ever before might simply not be able to take that step back and enjoy such a painting in the way we can, owing to our comparatively limited visual abilities. Who is the one with the disability here?

Two rejoinders to this brief set of examples come to mind immediately: Firstly, these might represent less sophisticated forms of posthuman enhancement – what we envision is someone who not only enjoys superior abilities but is also able to adjust them at will! More importantly, as the social model of disability would suggest, the problem is society's (and society's to fix), not that of the disabled-enhanced person. These are but teething pains in the journey towards a properly posthuman society, of the kind postulated by Bostrom (2008, 112):

> You have just celebrated your 170th birthday and you feel stronger than ever. Each day is a joy. You have invented entirely new art forms, which exploit the new kinds of cognitive capacities and sensibilities you have developed. You still listen to music – music that is to Mozart what Mozart is to bad Muzak. . . . You play a certain new kind of game which combines VR-mediated artistic expression, dance, humor, interpersonal dynamics, and various novel faculties and the emergent phenomena they make possible. . . . When you are playing this game with your friends, you feel how every fiber of your body and mind is stretched to its limit in the most creative and imaginative way.

If one was already convinced of the moral permissibility or even urgency of enhancement, then these rejoinders may well suffice. However, thinking with a social-model-of-disability frame of mind, one would wish to supply a sense of realism. Posthuman qualities often come proposed with a great deal of indeterminacy – we know neither the precise method for achieving them, nor the knock-on effects on other qualities. Nevertheless, going by what we already know, as Fitzgerald (2008, 42–43) points out extreme intelligence seems to correlate with sociopathic or psychopathic behaviour, for instance. Fitzgerald also warns that even if one were able to achieve a "partial effect" on slowing down the ageing process by genetic manipulation, it is no "guarantee that the extension [of lifespan] also provides for the capacity to engage in the kinds of activities one might desire for one's longer life."

Furthermore, given the speculative nature of many enhancement techniques, it is inconceivable at present that any such techniques can be refined without a lot of experimentation, including on human subjects (which conjures another set of ethical considerations). The subjects of such experimentation will have to bear the brunt of piecemeal enhancement which, I think, will more closely resemble the set of potential disabilities I outlined earlier than Bostrom's picture. Before they get to play anything close to the "certain new kind of game" of which Bostrom speaks, they will have to deal with a society whose social, economic, cultural, and physical structures are "fine-tuned" towards the normal range of human abilities.

These considerations point to a more general difficulty with posthuman speculation: By what criteria do we decide which attributes or capabilities to enhance? Any proposed programme of enhancement must be interrogated for its theoretical presuppositions. The social model of disability, after all, posits that disabled people are oppressed by society – and in particular by capitalist values (Anastasiou and Kauffman 2011, 371). Whether capitalist values are oppressive or not is a distinct question, but, at the risk of arguing simultaneously for two contrary theses, the point here is this: Does the choice of certain qualities as candidates for enhancement entrench values that are currently the causes of oppression themselves? Even if in the short term, those who have been enhanced might paradoxically become disabled, one should also take seriously the possibility in the long run that their presence may disrupt society by changing the balance of power between, say, age groups or people with different cognitive abilities (Clarke 2016, 220–21).

Conclusion

Disability therefore presents a challenge to posthuman enhancement – if one wishes to go beyond the human as we know her, how do we know whether what lies beyond is enhancement or disability? At what point might the tables be turned, as it were? If anything, this suggests that the "post" in "posthuman" is hardly a straightforward concept or aim.

No doubt, the social model of disability has been called into question for allegedly overstating the social roots of disability and for denying the importance of the body itself as a locus of disability (see, for example, Shakespeare and Watson 2001; Owens 2015). But this, it seems to me, strengthens rather than weakens the foregoing critique of enhancement. The social model of disability, it is thought, needs to better account for the interplay between society and biology (Anastasiou and Kauffman 2013,

450). Our discussion has been testament to just such an interplay. Just as social considerations of equity ought to influence our approach to allocating resources for enhancements in healthcare, whether certain biotechnological interventions are truly "enhancements" in the first place is a question that needs to be evaluated in the light of what new forms of disability and social oppression they might create. Only if we recognise that social structures develop through the often gradual mediation of biological factors will we appreciate how haphazard and experimental enhancement is likely to cause more grief than posthuman contentment.

Bibliography

Anastasiou, Dimitris, and James M. Kauffman. 2011. "A Social Constructionist Approach to Disability: Implications for Special Education." *Exceptional Children* 77(3): 367–84.

———. 2013. "The Social Model of Disability: Dichotomy Between Impairment and Disability." *The Journal of Medicine and Philosophy* 38(4): 441–59.

Austriaco, Nicanor Pier Giorgio. 2017. "Healthier than Healthy: The Moral Case for Therapeutic Enhancement." *The National Catholic Bioethics Quarterly* 17(1): 43–49.

Aydin, Ciano. 2017. "The Posthuman as Hollow Idol: A Nietzschean Critique of Human Enhancement." *The Journal of Medicine and Philosophy* 42(3): 304–27.

Bostrom, Nick. 2008. "Why I Want to Be a Posthuman When I Grow Up." In *Medical Enhancement and Posthumanity*, edited by Bert Gordijn and Ruth Chadwick, 107–36. Dordrecht: Springer.

Clarke, Steve. 2016. "Buchanan and the Conservative Argument against Human Enhancement from Biological and Social Harmony." In *The Ethics of Human Enhancement: Understanding the Debate*, edited by Steve Clarke, Julian Savulescu, Tony Coady, Alberto Giubilini, and Sagar Sanyal, 211–24. Oxford: Oxford University Press.

Cohen, Joshua T., and Peter J. Neumann. 2019. "The Cost Savings and Cost-Effectiveness of Clinical Preventive Care." *The Synthesis Project*, Research Synthesis Report No. 18. Robert Wood Johnson Foundation. Accessed 28 December 2019. www.rwjf.org/en/library/research/2009/09/cost-savings-and-cost-effectiveness-of-clinical-preventive-care.html.

Daniels, Norman. 2000. "Normal Functioning and the Treatment-Enhancement Distinction." *Cambridge Quarterly of Healthcare Ethics* 9(3): 309–22.

Eijk, Willem Jacobus Cardinal. 2017. "Is Medicine Losing Its Way? A Firm Foundation for Medicine as a Real *Therapeia*." *The Linacre Quarterly* 84(3): 208–19.

Fitzgerald, Kevin. 2008. "Medical Enhancement: A Destination of Technological, not Human, Betterment." In *Medical Enhancement and Posthumanity*, edited by Bert Gordijn and Ruth Chadwick, 39–53. Dordrecht: Springer.

Juengst, Eric T. 1997. "Can Enhancement Be Distinguished from Prevention in Genetic Medicine?" *The Journal of Medicine and Philosophy* 22(7): 125–42.

Owens, Janine. 2015. "Exploring the Critiques of the Social Model of Disability: The Transformative Possibility of Arendt's Notion of Power." *Sociology of Health & Illness* 37(3): 385–403.

Shakespeare, Tom, and Nicholas Watson. 2001. "The Social Model of Disability: An Outdated Ideology?" In *Exploring Theories and Expanding Methodologies: Where We Are and Where We Need to Go (Research in Social Science and Disability, Vol. 2)*, edited by Sharon N. Barnartt and Barbara M. Altman, 9–28. Bingley: Emerald Group Publishing.

2 Rethinking the Posthuman in Bioethics

David Boden and Sarah Chan

The "posthuman" in bioethics has been strongly linked to the discourse over human enhancement and the idea of radical species transformation produced by technological improvements on the human condition. In contrast to many other posthumanist perspectives, which aim to critically interrogate and de-centre the concept of "the human," bioethical accounts of the posthuman as the end-product of human enhancement *re*-centre the human, as simultaneously that which the posthuman goes beyond, and the prototype for function on which the posthuman improves. The figure of the posthuman as "human-plus" or "Humanity 2.0" provokes both excitement, from advocates of enhancement and transhumanists who embrace these visions of a hyper-technologised future, and disquiet, from those who oppose the changes they fear such a future might bring.

Although many of the technological possibilities that both sceptics and enthusiasts of posthuman enhancement envision remain hypothetical and speculative, these discourses in themselves nevertheless re-inscribe normative assumptions about the human condition and the category of "human." Such assumptions in turn may exert real and present influence, particularly within the sphere of health and biomedical policy and practice, in ways that demand further scrutiny. To this end, drawing on wider posthumanist perspectives that offer alternative possibilities for reconceptualising the human may provide scope to develop more open and critical renderings of the bioethical posthuman.

This chapter suggests the need to reimagine the posthuman in bioethics through a critical posthumanist lens. It does so first by providing an account of the figure of the posthuman, as it has arisen in bioethical discussions of human enhancement, as the desirable end-product of technology-mediated evolution. We show how the development of such arguments in the context of emerging medical technologies converged with transhumanist ideologies, leading to claims of a moral obligation to enhance. Drawing on perspectives from disability studies, we argue that this "posthuman

DOI: 10.4324/9781003020707-4

imperative" in fact reinforces the same normative values around human nature, function and embodiment that it purports to transcend. To counter this, we suggest that critical posthumanist approaches might offer scope for creative disruption of "human" norms, providing new perspectives to challenge the conventional framings of enhancement discourse. Finally, we turn to consider the implications of a "critical posthumanist turn" for policy. We briefly discuss the example of human heritable genome editing to illustrate how the values implicit in enhancement discourse can assume salience in a policy setting, and how critical posthuman reframings might or should reshape this.

Human, Transhuman, and Posthuman?

In mainstream bioethics, the "posthuman" has so far featured primarily as the highly technologised product of radical biomedical intervention. The sorts of biotechnologies that have been envisioned as the path to posthumanity span a diverse range: genetic technologies (Loftis 2005); pharmaceutical agents for cognitive, emotional and physical enhancement (Flower 2012); the integration of cybernetic prostheses (Barfield and Williams 2017) and more. While these emerging technologies have served as the impetus for taking seriously the prospect of enhancing human traits and functions, many still-speculative technologies are also central to bioethical accounts of the posthuman. Examples include nanorobotics and medical machines (Soto and Chrostowski 2018), anti-aging medicine (Juengst et al. 2003), and machine-brain interfaces (Lebedev and Nicolelis 2006). Together, both present and speculative technologies are imagined as part of an ongoing process whereby continued scientific progress will enable humanity to radically augment organic and inorganic life in a way hitherto unprecedented (Khushf 2007).

To understand why and how this hyper-technologised account of the bioethical posthuman has arisen, it is useful to view the evolution of bioethical posthumanist discourse from a historical perspective. Many of the key bioethical debates have arisen over reproductive and genetic technologies, such as *in vitro* fertilisation, pre-implantation genetic testing, and genetic modification. These technologies, capable of altering human functioning or affecting what sorts of humans might come into existence, were developed in the context of medicine aimed at addressing *dys*function. Initial bioethical questions therefore concerned their use in this context: whether or not it was permissible to use assisted reproductive technologies to allow otherwise medically infertile couples to have children,[1] genetic screening to enable parents to have healthy children, or genetic modification to treat disease (see, for example Glover 1984).

As these early technologies became more established, even as they continued to develop, the debate progressed to the possibility of applying these methods outside their original purposes – going "beyond therapy" (as per the President's Council on Bioethics 2003) to the realms of what became known as "human enhancement." Here, it converged with another discourse that had been emerging for some time in both popular and philosophical thoughts that of "transhumanism."

Transhumanism in one form or another is a part of humanity's cultural, social, and historical memory, even when not expressly named. From the Epic of Gilgamesh, which details a king's quest for immortality (see Bostrom 2005), to the rational humanism of Francis Bacon (see Saage 2013) and the rise of science fiction writing in the early twentieth century (Mirenayat et al. 2017), transhumanist aspirations and beliefs can be found in a multitude of different contexts. The origin of the term "transhumanism" and its modern ideology, however, can be traced back to Julian Huxley, who wrote:

> The human species can, if it wishes, transcend itself – not just sporadically, an individual here in one way, an individual there in another way, but in its entirety, as humanity. We need a name for this new belief. Perhaps transhumanism will serve: man remaining man, but transcending himself, by realizing new possibilities of and for his human nature.
>
> (Huxley 1968, 76)

While for Huxley transhumanism meant "superseding of humanity by virtue of technology as a purely human work, moving away from religion" (Del Aguila and Solana 2015, 504), modern transhumanists seek to transcend human *biological* limitations, perhaps to a point where we become "posthuman," undergoing "such radical changes that the result could only be regarded as a posthuman being, and no longer as a human being" (Gordijn and Chadwick 2008, 4).

Specific visions of the posthuman vary with their technological focus: anything from cybernetically enhanced humans (Coeckelbergh 2011) to mind-uploaded consciousness (Olson 2017). Nonetheless, these diverse visions share key ideological characteristics. While transhumanists envision a future where cyborgs, AI, social robots, animal-human chimeras and other diverse beings exist, these beings are perceived as a fulfilment of our species' inherent potential to alter the biological makeup of humanity through technological innovation. On this account, the posthuman is the ultimate endpoint of an ever-progressing co-evolution of humans and technology: from humans to enhanced humans to transhumans, and hence to posthumans (see for example Harris 2007; Chan 2008; Fuller 2011). Thus, although the posthuman is envisioned as a step-change from humanity as we now know it, the values underpinning this vision recapitulate, for the most part, liberal humanistic and predominantly individualistic ontologies. This has led to a somewhat self-contradictory stance whereby, for transhumanists, the posthuman is both a going beyond of the human while simultaneously reifying specific human qualities onto a "posthuman" face.

Arguments from evolutionary continuity notwithstanding, the superpower-equipped posthuman beings envisioned at the extremes of the discourse remain imaginary, as do most of the technologies that might create them. The posthuman, however, has nonetheless assumed considerable significance in discussions of contemporary human enhancements, as the imagined destination towards which these more quotidian interventions represent a first (or next) step. It plays a strong normative role within enhancement ethics narratives: either as an apocalyptic figure to be avoided, for those opposed to or sceptical of enhancement; or an aspirational one, to be sought, for those we might call "enhancement enthusiasts."

The concept of the posthuman in bioethics has thus been shaped, and remains constrained by, the process of its emergence: heavily biomedicalised and technology-focused, it tends to narrow and polarise debate. Taken in the historical context, this approach to the bioethical posthuman is understandable and has served to flatten the terrain for early policy discussions. We argue, however, that in confronting the imminent realities, rather than future imaginaries, of human biotechnological transformations, the utility of such caricatures of the posthuman to advance ethical and policy discourse is limited. Moreover, as we shall next argue, bioethical approaches to the posthuman often implicitly recapitulate biomedically based normativities in relation to human function, with potentially undesirable consequences.

Posthuman Nature: Reinforcing Normalcy?

As we have just seen, the bioethical posthuman has been positioned as the ultimate destination of a process of human evolution now mediated by technological as well as natural selection. This in turn implicates debates over human nature and the normative claims bound up with the concept. Human nature has been the basis of argumentative claims from both the pro- and anti-enhancement camps: those opposed to enhancement draw on an account of "human nature" that implies not being subject to our own processes of making, such that ambitions and attempts to re-engineer ourselves run counter to this nature or would destroy some essential part of it (see, for example Sandel 2009; see also President's Council on Bioethics 2003, 70; Meilaender 1997). Conversely, transhumanists and advocates of enhancement have argued that human nature is to be authors of our own creation, or to strive continually for greater achievement and self-improvement (see, for example Savulescu 2006; Chan 2008; Lawrence 2017). Re-interpreting human nature in the latter way is, perhaps, an understandable response to conservative stances that draw on the dichotomy between natural and artificial (where artificial indicates human-made) to suggest that technology is antithetical to (human) nature. As Hauskeller (2009) has pointed out, however, both sides of the argument nevertheless take certain things for granted by invoking human nature in this way; notably, both grant a normative force to "human nature" that it does not necessarily deserve.

Furthermore, there is an important difference between the possible accounts of human nature used in support of enhancement, of self-creators versus self-improvers. To be self-makers implies the exercise of agency and the availability of choices over our own fate and, given that most of the enhancements under discussion involve biomedical intervention on the human body, our own embodiment. It says nothing, however, about *how* that agency should be exercised, *which* choices should be preferred, and *what* the resulting forms of embodiment should be. The pursuit of continual improvement, on the other hand, has strong normative implications in relation to what constitutes improvement. Within typical pro-enhancement discourse, the narrative of human nature as constantly striving for betterment, and casting enhancement as a manifestation of this nature, is closely bound up with assumptions about which sorts of choices are better and which forms of embodiment are preferable.

This discourse is intrinsically problematic. As pro-enhancement arguments have converged with transhumanist ideologies, a characteristic step in establishing the acceptability of enhancements that go beyond the limits of normal human function has been to reject the moral significance of what is "normal" or "species-typical." Again, this responds partly to the position of those opposed to human enhancement who contend that therapy – that is, restoring normal function – is acceptable but enhancement, increasing function above normal, is not. Proponents of enhancement generally seek to overturn the distinction between therapy and enhancement and instead draw moral parallels between the two, arguing that both are beneficial to the recipient and hence something we have good reasons, perhaps even a moral obligation, to pursue (Savulescu 2005; Harris 2007; Chan and Harris 2007).

Yet, even while denying that "the normal" has any moral meaning in determining which interventions are permissible, such arguments tend to categorise certain states as dysfunction and others as enhancement or "hyperfunction." In doing so, though they reject the significance of "normal" as a threshold, they nevertheless accept implicitly the underlying assumption that "normal" is a point or a range on a linear, directional spectrum movement along which is necessarily worse in one direction, better in the other.[2]

Thus, even as medicalised and "ableist" approaches to disability have been criticised by Davis (1995) as "enforcing normalcy," the corresponding techno-medicalisation of enhancement *reinforces* the significance of normalcy and the evaluative hierarchy associated with it. By positioning normalcy as something to be transcended and setting a determinate direction for that process of transcendence, the transhuman/posthuman enhancement imperative in fact recapitulates existing hegemonic norms of human embodiment. While arguments supporting the permissibility of enhancement may have originated in liberal humanist ideals of freedom to pursue a "good life" by whatever means available, including technological, the arguments for an *obligation* to enhance tend towards the coercive in their normative visions of what a "good life" might entail and, especially, what the biomedical preconditions of such a life might be.

What Is Good for Us? Problems With the Posthuman Imperative

John Harris defines enhancement by stating, simply: "If it wasn't good for you, it wouldn't be enhancement" (Harris 2007). Framed in this way, the pro-enhancement imperative is hard to deny: how could what is "good for you" be anything other than good? Examining these arguments more closely, however, raises questions about the concept of the "good" that underpins them, particularly when contrasted with accounts of harms and the "harmed condition" that, Harris and others argue, we have a moral obligation to avoid (Harris 2001, 2005; Savulescu 2001). Framing disability primarily as a property of individual bodies and capacities, and using this as grounds for the moral imperative for bodily interventions on the basis of what is "good for you," imposes a firm hierarchy on different ways of being. In characterising disability and impairment as essentially "harmed" and harmful conditions, such claims impute value to certain forms of embodiment and disvalue to others, without scope for individuals' subjective determinations of what is good, or constitutes a good life, for them.

The putative "obligation to enhance" thus incorporates overly broad assumptions regarding objective well-being and the benefits of enhancement that in themselves embed certain implicit and problematic norms. Enhancement-as-obligation arguments start from the premise that there are some things which would almost certainly be bad for any being, such as extreme, unremitting pain and suffering, and draw the reasonable conclusion that it would be good for us to avoid these. The problem, however, occurs when these arguments depart from their initial terms to make much wider claims about what might be bad or good for us. Whether in relation to decisions for existing beings or choices about which sorts of beings should be brought into existence, arguments that invoke ideas of "a good life" or "the best possible life" to justify biomedical interventions imply both an objectively determinable standard of what makes a life good, and that the antecedents of a "good life" are primarily biologically or biomedically determined.

Again, this makes sense in terms of the narratives within which these arguments originated: the genetic and reproductive technologies among which they took shape were aimed initially at treating seriously life-limiting and life-threatening diseases. Such conditions, while not wholly inimical to a "good life," could be seen to constitute a significant impediment to well-being, such that most would agree one's life would go better without them, and that it would therefore be an acceptable use of technology to attempt to avoid them. From this starting point, however, the claim about acceptability based on beneficence and harm-avoidance undergoes an elision along two axes: in terms of the biomedical states-of-being to which it applies, where there is slippage to cover a vastly expanded range of potential indications; and in terms of the strength of the claim, from merely permissible to preferable or even obligatory.

We might view this extension partly as an argumentative device: in stating and defending a claim that some course of action might be so much morally better as to be *obligatory*, one clears the way to make it more plausibly *acceptable*, at least. Extending the argument in this way, however, also has the not-altogether-beneficial effect of narrowing its focus, to the point where, if it is to be considered for practical application, several relevant externalities have been overlooked. In particular, once we go beyond avoiding the sorts of conditions that are sufficiently bad that they would be universally undesirable (and the set of such conditions might be a much smaller one than we imagine), there is a great deal we could be doing to enable people to live better lives that does not involve these technologies at all.

This is of course not a novel argument, though as it provides principled support for neither a strong pro- nor anti-enhancement stance, it has received relatively little attention in the heretofore somewhat polarised debate (Almeida and Diogo 2019). Likewise, policy discussions have focused more on narrow questions of whether specific technological applications are permissible or not, rather than what might or should be done instead. By raising it here, however, we hope to demonstrate the need to move beyond reductionist for-and-against questions in relation to human enhancement. Instead, we should attend to the assumptions inherent in these questions themselves, and the ways in which asking them shapes the discourse.

"Good Lives" and Posthuman Flourishing

On the account of enhancement discussed earlier, disability and enhancement lie in opposite directions on a spectrum of function, such that "to fail to enhance is to disable." This framing, we have so far argued, incorporates two problematic assumptions: an account of disability and enhancement as solely or primarily linked to biomedical functioning; and an objective definition of the good whereby disability necessarily equates to harm and enhancement to benefit. Such an approach suffers from what Hughes has described as "the perceptual pathology of non-disablement. It is pathological because it is not neutral and because it thinks of itself as being so" (Hughes 1999, 164). In doing so, it forecloses alternative understandings of what sorts of "good lives" might be possible, different forms of embodiment and capacities notwithstanding.

Moreover, attempting to capture the complexity of what makes a "good life" or what is "good for one" merely in terms of biological function, beyond a certain very minimal required threshold, is clearly over-reductionist. Recognising this, many scholars have proposed more nuanced approaches that divorce the concept of enhancement from simple increases in functioning (Earp et al. 2014), characterising it more broadly in terms of increased welfare (Savulescu, Sandberg, and Kahane 2011) or well-being (Hofmann 2017).

The problem here, as Hauskeller (2009) points out, is that the relationship between subjective well-being and objective capacities is notoriously difficult to define: *pace* Mill's "pig satisfied," some things we think ought to be good for us, such as increased intelligence, do not necessarily conduce to greater subjective happiness. Advocates of enhancement, he notes, therefore often turn to other concepts such as "flourishing" in an attempt to circumvent this problem.

It seems reasonable to think that, all other things being equal, it is better for a being to have a greater rather than lesser capacity for flourishing where that capacity has a reasonable chance of being fulfilled. But what is "flourishing," and how are our criteria for what constitutes more or less flourishing determined? Hauskeller argues that such attempts tend, perhaps inevitably, to revert to some normative ideas of human nature to ground their account of flourishing. To illustrate this, he suggests that (for example) proponents of enhancement would probably hold that "reading and being able to understand and appreciate Proust is better than reading Tom Clancy" (Hauskeller 2009, 18). As he goes on to observe, however: "Most people do in fact prefer Tom Clancy to Proust, but we like to think that they *ought* to like Proust better because that would be more befitting to humans" (Hauskeller 2009, 18).

In other words, such understandings of enhancement as increased flourishing still rely on notions of the species-typical to define what it is for us to flourish *as* humans. This poses an additional problem for enthusiasts of posthuman enhancement, who seek to reject species-typical accounts of "normal" functioning as limiting what is proper or "good for us" as humans or *how much* of "the good" we should strive to attain, while still drawing on the species-typical to define *what* "the good" is that we should increase, what it is we should have more of.

The question of whether any remotely universalisable, species-neutral account of flourishing is possible (and if it is not, what should guide our decisions) is difficult, and one that we do not propose to solve here. Critical posthumanism, however, might prompt us to reflect on what (and who) has shaped these apparently axiomatic ideas of what constitutes "human flourishing" or a "good life": *why* do we think it must be better to appreciate Proust than Clancy? Whose worldview and experiences are being given primacy through this evaluative framework – and whose are being excluded? Likewise, defining disability as a "harmed condition" that "we have . . . a strong rational preference not to be in," or "that constitute[s] a harm to the individual, which a rational person would wish to be without" (Harris 2000, 97–98) should prompt us to ask: who are the "we" expressing this preference? What are the implications of classifying as *irrational* those people who are in such a condition and yet assert their satisfaction with being so? Questions about enhancement that are founded on such definitions may miss the opportunity and the need to interrogate more critically the definitions themselves. Again, alternative approaches to posthumanism might help us to surface and articulate broader concerns around how these framings of posthuman enhancement work to constrain and reinforce ideas (and ideals) of "the human."

Rethinking the Bioethical Posthuman

As Miah (2008) observes in his history of posthumanism across disciplines, bioethical posthumanism has mainly focused on the ethics of biomedical enhancement, gathering up Enlightenment-style aspirations along the way. The value of rethinking bioethical approaches to enhancement through a critical posthumanist lens is that it provides a call to refocus our enquiries; to pull back from the narrow terrain in which this debate has so far taken place and consider wider framings of the question; and to consider what we may have missed or minimised in our constructions of the issue so far. In particular, it affords us an avenue to attend to the social as well as the technological dimensions of enhancement, and to develop new paradigms such as social models of not only disability (Shakespeare 2013) but also enhancement (Chan 2020; see also Wee, this volume) that enable such attention.

We therefore suggest that there is a need to develop new understandings of the bioethical posthuman that have the potential to creatively disrupt existing norms and notions of "the human." This disruption may take place, not as enhancement critics may fear by rendering humans obsolescent (Sparrow 2019), but by challenging the hegemony of the "normal" body, which pro-enhancement discourses also embed through their positioning of the human as the subject of improvement. The dissolution of the human subject proposed by critical posthumanism may thus also serve to uncover the implicit and often unacknowledged normative biases of bioethical posthumanism.

In contrast to the transhumanist focus on science, technology, and reason in bringing about the posthuman condition, critical posthumanism focuses on reconceptualising the human, rather than using technology to transcend it. Seeking to break with the dominant assumptions and trends of Western ethics and ontology, particularly anthropocentric worldviews and liberal-humanistic conceptions of the self, critical posthumanists emphasise systems, such as human beings, and their entanglement with other subjects, objects, entities, and processes. For instance, Donna Haraway (1991) sees the cyborg as a destabilising metaphor seeking to undermine dualisms between nature and culture, human and nonhuman, artificial and organic and so forth. The posthuman on these accounts is an emergent ontology, one characterised by an unstable identity and shifting relationships. While technology serves as one impetus for re-conceptualising the place of humanity, critical posthumanism does not exclusively see technology as the guarantor of "posthumanity" (Gladden 2018, 46). Nonetheless, it can provide useful conceptual tools and framings with which to engage with the effects of ongoing technologisation, not just in re-shaping humans themselves, but the complex network of environments and other beings within which they are situated.

To further illustrate how a critical posthumanist approach might provide new insight into bioethical debates over enhancement, consider the concern that the possibility of human enhancement may render us "always already disabled," by casting the normal human condition as essentially pathological and the appropriate target of technological interventions. As an argument against enhancement, this can be read as comprising two linked claims: 1) that in our pursuit of unending technological fixes for our existence, we may lose sight of what is actually valuable about that existence; and 2) that in viewing the state of "human being" as inherently pathological, we disvalue something we ought to value. But these claims hold only if we regard being "disabled" as something that requires to be fixed via technology, or as a state that is intrinsically rooted in biomedical pathology. Seeing the human as perpetually *open* to the possibilities of technological intervention does not in itself constitute a normative statement about the desirability nor the obligatoriness of certain possibilities over others.

Insofar as the debate over posthuman enhancement has been characterised as a matter of "them" and "us," of deciding who should count as "us" and which are the "bodies that matter" (Zylinska 2004), a critical posthumanist perspective should encourage us to widen our conception of "us" to encompass not only the transhuman and bioethical idealised posthuman, but also other forms of difference, be they in terms of embodiment, functioning, species membership or something else. Humanity has often been understood by reference to what it has been historically defined in opposition to, whether nonhuman animals, monsters, aliens, or anything traditionally seen as the "other" to humanity (Gladden 2018, 47); critical posthumanism challenges exclusionary definitions of humanity, along with their anthropocentric implications.

As such, the "post" in posthuman is not so much a reference to beings with superhuman capacities and functions, but signifies a break with a particular view of human beings as individualistic, morally superior, autonomous beings who are ontologically separate or prior to the world they inhabit. Critical posthumanism aims to refashion the limits of humanity by replacing it with a post-humanity which includes biological, artificial, and hybrid agents (Gladden 2018, 48). A previous argumentative strategy to support the acceptability of future (post)human enhancements has been to emphasise their continuity with our existing relationship to technology, through the claim that "we are already posthuman," in terms of our self-constitution via technology. In our proposed new bioethical posthumanism, what is transformed is not human beings themselves, but the concept of humanity and its normative characteristics: the key step is not claiming that *we* ourselves are posthuman, but the (re)constitution of "the posthuman" as a moral species, together with what sorts of beings we recognise as belonging to that species, and what relationalities, responsibilities and obligations being posthuman implies.

Policy for Posthumans

Bioethics is often characterised in terms of its potential application: that it should be normative, action-guiding, and grounded in real-world context in order to help us determine what is best to do. This has a clear relevance when it comes to the potential of bioethics to help inform policy. It also, however, means that we should attend to the impact of the discourses we promulgate in the real world, even where the outcomes do not have an immediate bearing on policy. Thus, although most of the technologies for radical posthuman enhancement are not yet in existence and therefore not the subject of policy-making at present, the bioethical discussion surrounding these imagined technologies can nonetheless influence public discourse and attitudes and affect the direction of current policy debates – both in terms of the arguments made, and, perhaps more subtly, how the questions are framed.

One area of policy-making currently and prominently at stake is the regulation of human heritable genome editing, that is, human germline genetic modification, albeit by a somewhat new name. Genetic modification is of course the theoretical terrain on which, as discussed, much of the discourse over (post)human enhancement has been contested. The advent of genome editing technologies abruptly shifted these possibilities from theoretical to practical and revived this debate: the prospect of heritable genetic enhancement was suddenly real, requiring renewed attention. While previous decades and successive iterations of debate over which these arguments have played out have provided a well-established ethical and policy framework for regulation, genome editing may provide an opportunity to revisit these discourses with a critical posthumanist eye, and to identify and interrogate the ethical assumptions that underpin our current regulatory approaches to genetic and reproductive technologies.

Proponents of obligations to enhancement and procreative beneficence have generally stopped short of suggesting that these obligations should be enforced by policy. When it comes to the regulation of reproductive choices, a common approach has been to draw a distinction between what is morally obligatory and what should be legally obligatory: procreative beneficence may impose a moral obligation to choose the child with the best chance of the best possible life, but procreative liberty requires us to refrain from imposing a legal obligation to do so. Nevertheless, the way in which the discourse unfolded has, as argued earlier, produced a value framework within which some sorts of lives and forms of embodiment are strongly favoured over others. This has been recapitulated partly in law, at least in the UK, where much of this discussion has taken place and which has one of the more highly developed regulatory systems for such technologies. There may be no legal obligation to "enhance," but the regulatory framework surrounding reproductive choice reflects a marked preference, tending in some cases towards an obligation, not to "disable" – where both "enhancement" and "disability," as discussed, embed unjustified normative assumptions regarding preferable forms of embodiment.[3]

The opportunity offered by genome editing policy, then, is rather than simply recapitulating the same arguments that have characterised previous discussions, to take a different approach that might open up new ways of thinking about technology, human and posthuman bodies and the relationships between them. We suggest that a posthumanist perspective on enhancement technologies should be, not merely liberal regarding what choices we are *free* to make, but agnostic regarding a much wider range of those choices as to what decision would be *better* to make. This may require us to take a step back from the strong pro-technology position that relies, as it has for decades, on the intuitive and easily argued imperative to cure disease, alleviate suffering and prevent harm; and to acknowledge that some of the differences that an over-extension of this argument has cast as "harms" may not, in fact, be so dreadful after all. It is not necessary to assert a moral obligation to enhance in order to defend the moral permissibility of a wide range of choices about how to be and live, whether classed as therapy, enhancement, disability, or otherwise. Furthermore, we should recognise that the enhanced posthuman as futuristic technological imaginary is of limited use in advancing present-day discourse over current technological realities. Arguments about the sorts of genetic enhancements that might be obligatory under hypothetical conditions of genome editing being safe, effective and equitably available do not help us, in the here and now, to chart a course towards a future where these conditions are met (Chan 2019).

Policy for posthumans demands that we recognise and respond to the varied needs, perspectives and (inter)subjectivities of our *contemporary* posthuman community, far more than the threats and promises of future imaginaries. To do so will require framing new debates as well as critically re-evaluating old ones, in terms that can recognise and account for the diversity, present and future, of our moral species.

Notes

1. Our use here of biomedicalised, heteronormative wording is deliberate and designed to illustrate that the normalcy enforced by medicine is not only biomedical: the construction of the "medically infertile couple" as the intended users of reproductive technology, particularly when contrasted with the term "social infertility" that was in the past applied to single women and same-sex couples, aptly demonstrates this!
2. This critique is related to but goes beyond analyses of the conceptual difficulties of "more is better" (Earp et al. 2014; Hofmann 2017): the problem as we see it is not only *whether* more or less of a given function is better, but *that* a particular direction of travel can be said to be better at all.
3. Consider the stipulation that when selecting embryos, those at risk of "a serious physical or mental disability . . . must not be preferred to those that are not known to have such an abnormality" (Human Fertilisation and Embryology Act 1990 s13(9)); this effectively constitutes a legal prohibition against particular reproductive choices. Even where the law is permissive rather than directive (as for example in the case of abortion law, under which one of the justifications for otherwise-illegal late-term abortion is the child having "such physical or mental abnormalities as to be seriously handicapped"), the permissibility of

certain choices versus others – aborting a disabled child is legal where aborting a "normal" child is not – indicates the underlying hierarchy of values attached to different lives.

Bibliography

Almeida, Mara, and Rui Diogo. 2019. "Human Enhancement: Genetic Engineering and Evolution." *Evolution, Medicine, and Public Health* 2019(1): 183–89. doi:10.1093/emph/eoz026.

Barfield, Woodrow, and Alexander Williams. 2017. "Cyborgs and Enhancement Technology." *Philosophies* 2(1): 4.

Bostrom, Nick. 2005. "A History of Transhumanist Thought." *Journal of Evolution and Technology* 14(1):1–25.

Chan, S. 2008. "Humanity 2.0? Enhancement, Evolution and the Possible Futures of Humanity." *EMBO Rep* 9 (Suppl 1): S70–S74.

———. 2019. "Commentary on 'Moral Reasons to Edit the Human Genome': This Is Not the Moral Imperative We Are Looking for." *Journal of Medical Ethics* 45(8): 528. doi:10.1136/medethics-2018-105316.

———. 2020. "Therapy, Enhancement and the Posthuman." In *The Bloomsbury Handbook of Posthumanism*, edited by Mads Rosendahl Thomsen and Jacob Wamberg, 215–30. London: Bloomsbury Academic.

Chan, S., and J. Harris. 2007. "In Support of Enhancement." *Studies in Ethics, Law and Technology* 1(1): Article 10.

Coeckelbergh, Mark. 2011. "Vulnerable Cyborgs: Learning to Live With our Dragons." *Journal of Evolution and Technology* 22(1): 1–9.

Davis, Lennard. 1995. *Enforcing Normalcy: Disability, Deafness and the Body*. London: Verso.

Del Aguila, Jorge Walker Vásquez, and Elena Postigo Solana. 2015. "Transhumanismo, neuroética y persona humana." *Revista Bioética* 23: 505–12.

Earp, Brian. D., Anders Sandberg, Guy Kahane, and Julian Savulescu. 2014. "When Is Diminishment a Form of Enhancement? Rethinking the Enhancement Debate in Biomedical Ethics." *Front Syst Neurosci* 8: 12. doi:10.3389/fnsys.2014.00012.

Flower, Rod. 2012. "The Osler Lecture 2012 'Pharmacology 2.0, Medicines, Drugs and Human Enhancement'." *QJM* 105(9): 823–30.

Fuller, Steve. 2011. *Humanity 2.0: What It Means to Be Human Past, Present and Future*. Basingstoke: Palgrave Macmillan.

Gladden, Matthew E. 2018. *Sapient Circuits and Digitalized Flesh: The Organization as Locus of Technological Posthumanization*. Indianapolis, IN: Defragmenter Media.

Glover, Jonathan. 1984. *What Sort of People Should There Be?* New York: Penguin.

Gordijn, Bert, and Ruth Chadwick. 2008. "Introduction." In *Medical Enhancement and Posthumanity*, edited by Bert Gordijn and Ruth Chadwick. Dordrecht: Springer.

Haraway, Donna. 1991. *Simians, Cyborgs and Women: The Reinvention of Nature*. New York: Routledge.

Harris, J. 2000. "Is There a Coherent Social Conception of Disability?" *Journal of Medical Ethics* 26(2): 95–100.

———. 2001. "One Principle and Three Fallacies of Disability Studies." *Journal of Medical Ethics* 27(6): 383–87.

———. 2005. "Reproductive Liberty, Disease and Disability." *Reprod Biomed Online* 10(Suppl 1): 13–16.

———. 2007. *Enhancing Evolution*. Princeton: Princeton University Press.

Hauskeller, Michael. 2009. "Prometheus Unbound: Transhumanist Arguments From (Human) Nature." *Ethical Perspectives* 16(1): 3–20.

Hofmann, Bjørn. 2017. "Limits to Human Enhancement: Nature, Disease, Therapy or Betterment?" *BMC Medical Ethics* 18.

Hughes, Bill. 1999. "The Constitution of Impairment: Modernity and the Aesthetic of Oppression." *Disability & Society* 14(2): 155–72. doi:10.1080/09687599926244.

Huxley, Julian. 1968. "Transhumanism." *Journal of Humanistic Psychology* 9(1):73–76.

Juengst, Eric T., Robert H. Binstock, Maxwell Mehlman, Stephen G. Post, and Peter Whitehouse. 2003. "Biogerontology, 'Anti-aging Medicine,' and the Challenges of Human Enhancement." *Hastings Center Report* 33(4): 21–30. doi:https://doi.org/10.2307/3528377.

Khushf, George. 2007. "The Ethics of NBIC Convergence." *Journal of Medicine and Philosophy* 32(3): 185–96. doi:10.1080/03605310701396950.

Lawrence, David. 2017. "More Human than Human." *Cambridge Quarterly of Healthcare Ethics* 26(3): 476–90. doi:10.1017/S0963180116001158.

Lebedev, Mikhail A, and Miguel A. L Nicolelis. 2006. "Brain-Machine Interfaces: Past, Present and Future." *Trends in Neurosciences* 29(9): 536–46.

Loftis, J. Robert. 2005. "Germ-Line Enhancement of Humans and Non-Humans." *Kennedy Institute of Ethics Journal* 15(1): 57–76.

Meilaender, G. 1997. "Begetting and Cloning." *First Things* 74(41–43). doi:D – KIE: 60102 OTO – KIE.

Miah, Andy. 2008. "A Critical History of Posthumanism." In *Medical Enhancement and Posthumanity*, edited by Bert Gordijn and Ruth Chadwick, 71–94. Netherlands: Springer.

Mirenayat, Sayyed Ali, Ida Baizura Bahar, Rosli Talif, and Manimangai Mani. 2017. "Science Fiction and Future Human: Cyborg, Transhuman and Posthuman." *Theoretical and Applied Linguistics* 3(1): 76–81.

Olson, Eric T. 2017. "The Central Dogma of Transhumanism." In *Perspectives on the Self*, edited by Boran Berčić, 35–57. Rijeka: University of Rijeka.

President's Council on Bioethics. 2003. *Beyond Therapy: Biotechnology and the Pursuit of Happiness*. Washington, DC: President's Council on Bioethics.

Saage, Richard. 2013. "New Man in Utopian and Transhumanist Perspective." *European Journal of Futures Research* 1(1): 14. doi:10.1007/s40309-013-0014-5.

Sandel, Michael. 2009. *The Case Against Perfection: Ethics in the Age of Genetic Engineering*. Harvard: Harvard University Press.

Savulescu, Julian. 2001. "Procreative Beneficence: Why We Should Select the Best Children." *Bioethics* 15(5–6): 413–26.

———. 2005. "New Breeds of Humans: The Moral Obligation to Enhance." *Reprod Biomed Online* 10(Suppl 1): 36–39.

————. 2006. "Justice, Fairness, and Enhancement." *Annals of the New York Academy of Sciences* 1093: 321–38.

Savulescu, Julian, Anders Sandberg, and Guy Kahane. 2011. "Well-Being and Enhancement." In *Enhancing Human Capacities*, edited by Julian Savulescu, Ruud ter Meulen, and Guy Kahane. Oxford: Wiley-Blackwell.

Shakespeare, Tom. 2013. "The Social Model of Disability." In *The Disability Studies Reader*, edited by Lennard Davis, 214–21. New York: Routledge.

Soto, Fernando, and Robert Chrostowski. 2018. "Frontiers of Medical Micro/ Nanorobotics: In Vivo Applications and Commercialization Perspectives Toward Clinical Uses." *Frontiers in Bioengineering and Biotechnology* 6(170). doi:10.3389/fbioe.2018.00170.

Sparrow, Robert. 2019. "Yesterday's Child: How Gene Editing for Enhancement Will Produce Obsolescence – and Why It Matters." *The American Journal of Bioethics* 19(7): 6–15. doi:10.1080/15265161.2019.1618943.

Zylinska, Joanna. 2004. "The Universal Acts." *Cultural Studies* 18(4): 523–37. doi: 10.1080/0950238042000181647.

3 Gen-Ethics and the Posthumanities

Ruth Chadwick

At first sight, it might seem that ethics related to the human genome is firmly *pre*-"posthumanities." It is, after all, focused on the human and thus anthropocentric. I shall suggest, however, that there are reasons for thinking otherwise. There are at least four ways in which ethical issues in this field have widened – in particular, beyond a focus on and particular understanding of the individual human subject, and indeed, these four ways can be seen as representing a gradual broadening of concern. These include, at least in some areas, a move away from individual-centred ethics to a focus on how to promote the public good and the sharing of the benefits of science; the recognition of the similarities between the human genome and that of other species, implying a move beyond anthropocentrism; the increasing realisation of complexity and the importance of other "omics" beyond genomics, requiring a greater awareness of the environmental situatedness of the human; and the developing discussion of gene editing, along with consideration of human and transhuman (which for present purposes means bio-enhanced human) futures.

The Human Genome, Individual, and Collective

The human genome has acquired an iconic status. It has been described as the "common heritage of humanity" (see, e.g. UNESCO 1997). This may be used as the basis for an argument against genetic modification of the germline such as germline gene therapy or gene editing. The very fact that the language of common heritage has been used is arguably itself suggestive of a distancing of the issues from the interests of individuals. The Human Genome Project (henceforth HGP) aimed to produce the human genomic sequence but not the sequence of any given individual (see Lander et al. 2001). The HGP had a significant impact upon ethical debate in several ways. There was a financial effect, via the dedication of part of the budget to the consideration of ethical, legal, and social issues. Beyond that, however,

DOI: 10.4324/9781003020707-5

the HGP and subsequent genomic research had an influence on how ethical issues have come to be framed. The Human Genome Organisation (henceforth HUGO), established to coordinate efforts in human genome sequencing, itself set up an Ethics Committee with an international and multidisciplinary membership.

To a considerable extent, the work of the Ethics Committee of the Human Genome Organisation provides a useful case study of the ways in which gen-ethics has changed and developed over time. The earliest Statement issued by the Committee (HUGO 1996) presented ten "commandments" or "Cs" (because they all involved a concept beginning with a "C") including such well-recognised principles as competence, consent, confidentiality, compensation, and consultation. It is not inappropriate, then, to regard this Statement as enunciating widely recognised standards for the conduct of genomic research. As such they did indeed reflect, primarily, the individual-focused bioethics that was prominent in much of the second half of the twentieth century. The Statement on DNA Sampling which followed soon after was explicit about the fact that individual interests may sometimes need to be secondary:

> These shared biological risks create special interests and moral obligations with respect to access, storage and destruction that may occasionally outweigh individual wishes.
>
> (HUGO 1998)

After the completion of the HGP at the turn of the century, the genomics community turned to new challenges, and the HUGO along with it. The sequencing of the genome had implications both for the role that genes play in human life and for humans' self-perception. The HGP revealed that the number of genes in the human genome was significantly fewer than previously thought, and this led to the recognition of levels of complexity in the interaction between genetic and other factors. As regards self-perception, any idea of a simple relationship between genes and personal identity was shown to be untenable. Both genetic essentialism and genetic exceptionalism appeared to be undermined. That did not mean a disappearance of interest in what information our genes *can* provide: there remains a flourishing market for direct to consumer genetic testing, and thus an appetite for what information this testing can provide about our ancestry and health status.

In this phase, and over time, the Ethics Committee developed a particular ethical approach, based on the principles of solidarity and equity, as evidenced in its Statements on Benefit-Sharing (2000), on Genomic Databases as Global Public Goods (2002), and on Pharmacogenomics (HUGO Ethics Committee 2007). The emphasis marked a move beyond an individualistic focus and a recognition of the interests of relatives towards sharing what benefits might accrue from genomic research globally, including to underserved populations. To a certain extent this might be seen as an aspect of what has been called the communitarian turn in bioethics generally in the 1990s (see Chadwick 1998), but arguably also it reflects the realisation that emerging issues in genomics needed fresh thought. The now classic example of this is that of informed consent in the context of biobanks and population genomic research (see, e.g. Chadwick and Berg 2001). This type of research represented a significant shift away from conventional biomedical research, such as pharmaceutical trials, which required specific consent to participate in relation to the testing of a particular drug. The turn towards broad consent to donate samples to a biobank and for those samples to be used for research was linked with (still ongoing) arguments for the public good. In order to achieve purported public health benefits in the long term, large-scale collections of samples and big data affording significant statistical power were said to be needed. The work of the Ethics Committee in this phase can be viewed against this background.

In 2010, the Ethics Committee was replaced by a new committee, the Committee on Ethics, Law and Society, and this marked another change of direction. Although the Ethics Committee had included different disciplines, this renaming and reconstitution emphasised the role of those disciplines. The outputs of the new committee tended to be papers rather than Statements, including papers on gene editing and the public good (Capps et al. 2017) and on gene editing through the lens of solidarity (Mulvihill et al. 2017). The new Committee, whose membership had some continuity with its predecessor, thus continued the ethical stance of solidarity and equity, and the focus on the possibility of harnessing the outcomes of scientific research for public good.

To sum up this first point, then, gen-ethics had an impact on bioethics through its emphasis on the shared nature of the human genome and the ideas about capturing the benefits of science for human populations broadly speaking, rather than for the few. So the emphasis, as far as this is considered, remains human, but with a move away from the centrality of the individual human subject which for so long was prominent in much humanities thinking, and remains so in some contexts – for example, the autonomy of the individual patient retains its importance in medical ethics and law.

Relationships With Other Species

Developments beyond this, however, were required by other aspects of the outcome of the HGP. The 2001 Lander et al. paper on the initial sequencing and analysis of the human genome already noted the importance of comparative genomics (Lander et al. 2001). It became clear not only that the number of genes in the human genome was smaller than expected, but also that there was a (surprising to some, perhaps) extent of similarity between the human genome and the genomes of other species. The question arises as to what the implications of this are. In an article in *Nature*, Waterson et al. wrote:

> Our close biological relatedness to chimpanzees not only allows unique insights into human biology, it also creates ethical obligations. . . . We hope that elaborating how few differences separate our species will broaden recognition of our duty to these extraordinary primates that stand as our siblings in the family of life.
>
> (Waterson, Lander, and Wilson 2005)

The significant point here is that Waterson et al. do argue that the genomic similarities create ethical obligations. Against this it might be argued that facts do not, of course, determine our ethical obligations. We can take different positions in relation to the same facts. However, while debate may continue about the precise ethical implications (on which more details are given in the following), the extent of relatedness is given and this inevitably has an influence on human self-perception in relation to other life forms.

The recognition of the human position in relation to other animals has led some ethicists to go beyond the public good arguments outlined in the previous section. For example, Capps and Lederman write about the concept of "One Health" defined as "an integrative and interdisciplinary effort to improve the lives and well-being of human beings and nonhuman animals, as well as to preserve the environment":

> we consider the concept of public biobanking as promoting the public good, and suggest that there is potential to expand this concept to universal goods: benefits that plausibly extend beyond human interests, to those of animals, and even further to include environmental conservation.
>
> (Capps and Lederman 2015, 258)

Capps and Lederman here point out that the threat of emerging zoonotic diseases has led to the idea of One Health gaining traction (ibid., 261). The Covid-2019 pandemic may reinforce this trend, but genomic similarities are another relevant factor to be taken into account.

Epigenomics: Humans and the Environment

Widening the focus further, much has been and continues to be said about the ways in which our microbiome contributes to our health and life (see Le Page 2020). The very boundaries of the human are being rethought (not only in relation to microbes but also as regards the digital and artificial intelligence). I would argue that it is also important to recognise that epigenomics requires us to think about the human genome, and indeed the human, in a different way. It demands a recognition of the "situatedness" of humans in ways that go beyond both our relationships with other species and even the concepts of relational autonomy and solidarity that have been promoted in those approaches to ethics that have consciously moved away from the centrality of the isolated autonomous individual self, such as feminist ethics and communitarianism.

Epigenomics, literally what is "over and above" the genome, deals with the factors which affect the ways in which genes are expressed, by switching them on or off. Of the mechanisms of epigenetics, the most well-known is methylation, whereby a methyl group attaches to the genome at specific points. It is established that environmental factors and lifestyle can bring about methylation changes in the genome, and this has led to consideration of the implications for responsibilities in pregnancy, for example, suggesting that women should have an eye to the possible consequences of their behaviour for epigenetic changes in their offspring. This has not surprisingly provoked some responses showing that male behaviour can also have an effect (see Richardson et al. 2014). Epigenetic changes in their sperm may follow exposure to toxins over their lifetime. Whole communities may show marked differences in methylation patterns, reflecting different styles of life, for example, between farmworkers and non-farmworkers (Howard et al. 2016).

The upshot of this is that epigenomics provides a clear demonstration of the ways in which the genome interacts with the environment in ways that may be to a large extent beyond our control. Stress, for example, can produce epigenetic changes that appear to be transmissible across generations. Thus, second-generation migrants have been shown to have a higher incidence of poor mental health than the indigenous population (Tahira and Agius 2012).

So, the human genome has to be considered in context. Moreover, the individual human being must be considered in relation to their environment. There are inevitable ethical and self-perception implications here. Chiapperino and Testa suggest that "different framings of epigenomic evidence and empowerment discourses . . . are in fact likely to have a fundamental bearing upon the roles and obligations of agents in the emerging vision of

PM [personalised medicine], including quite possibly on our self-understanding as citizens, patients and health-care consumers" (Chiapperino and Testa, 2017).

Even in medical ethics, then, the picture of the autonomous individual taking decisions about their own body is too narrow to cope with epigenomics, in a situation where decisions about one's lifestyle and environment can affect future generations through epigenomic mechanisms (Chadwick 2017). There are difficult questions about responsibility for the next and for future generations, which involve taking into account much more information than previously supposed.

Gene Editing: Human Futures

While epigenomics directs human attention to the environment and to future generations, this again does not go far enough for some. Considering the human species in relation to its environment may provide a reason to want to change humans. In the face of climate change, for example, it has been argued that it may be necessary to edit the human genome in order to facilitate moving to other environments, other planets. Thus, John Harris says:

> The problem is that progress via Darwinian evolution is extremely slow, and the direction unpredictable; all we know is that it will facilitate gene survival. It is probable that, in the interests of human survival and certainly those of human welfare and well-being, we may simply not be able to wait. For example, we will need to accelerate the development of better resistance to bacteria, disease, viruses or hostile environments or of the technologies that will be eventually necessary to find, and travel to, habitats alternative to the earth.
>
> (Harris 2016)

As stated earlier, the genome as the common heritage of humanity may be regarded as a reason against interfering in the human germline, and thus there has been an oft-cited distinction between somatic and germline therapy. The same distinction has been drawn in relation to gene editing. The case of He Jiankui in China provoked a shocked response when it emerged that a germline editing experiment had led to the birth of twins who had been modified in an attempt to make them immune to HIV. Prior to that there had been a widespread, though not universal, consensus against clinical use of germline editing, although it is arguable that there is increasing, if gradual, softening of what was once treated as a firm line.

Some have suggested that He Jiankui's experiment was problematic ethically, not because it involved the germline but because other principles of research ethics were not followed, or not followed correctly, such as informed consent (see, e.g. Savulescu and Singer 2019). So Savulescu and Singer propose an ethical pathway towards clinical translation, starting with catastrophic and then severe genetic disorders as candidates for intervention. Others go further and ask why, even if it is thought that there is something special about the human genome, it should be preserved in its *current* state (Harris 2016).

Of course, not all of those who support germline editing do so on the grounds of preventing disease, or responding to threats such as climate change. If the bar to editing the germline were to be lifted, that would open the way, in principle, to transforming humans in ways that might make them no longer indisputably human. Once again, the boundaries of the human are in dispute (see, e.g. Hughes et al. 2007). It is notable that Harris talks about what is important for *human* survival, while others envisage a *trans*human future of bioenhanced beings. Rob Sparrow, on the other hand, has written about how enhancements may lead to a built-in obsolescence, as new and "improved" versions are created (Sparrow 2019).

Thus gen-ethics has prompted, or at least supported, a widening of bioethics in a number of ways, an acknowledgement of the posthuman move beyond anthropocentrism. Beginning with a focus on sharing the benefits of science with humanity as a whole, rather than focusing on an individual-human-centred ethics, it has contributed to a necessary realisation of human situatedness in the context of other species and the environment. This involves rethinking the boundaries and nature of the human. Finally, the prospect of gene editing gives serious bite to the prospect of human enhancement and the possibility of a bioenhanced transhuman future. It is now pertinent to consider some possible caveats.

Caveats and Counterarguments

Although gen-ethics has developed in interesting ways that may indicate at least a small turn towards posthumanist thinking, it might be argued that it nevertheless remains firmly welded to a position which emphasises the importance and centrality not only of the human but also of the individual human subject in ethics.

In ethics itself, the concern for public good which has so marked the development of gen-ethics in the past 20 years at least, has not *replaced* the concern for the individual subject, most notably in relation to issues such as informed consent and data privacy. Although the concept of broad consent, for example, has been widely accepted in the context of biobanks, it remains the case that it is individuals who give that consent, and specific informed consent is still needed in other contexts such as clinical trials. Nevertheless, there has been recognition of wider considerations, representing an ongoing tension between individual and community interests.

Secondly, it might be argued that the recognition of our relationship to other species has had a negligible effect on the development of ethics, despite the recommendation in the *Nature* article described earlier. What are the possible reasons for this? There are several aspects to the relationship between biological facts and inter-species relations. As was suggested earlier, it is possible to draw different conclusions from our close relatedness to other species. One possibility, which is alluded to in the *Nature* article (Waterson, Lander, and Wilson 2005), is to argue that the similarity makes those other species particularly well suited as experimental models for the benefit of human beings. This is very different from humans acknowledging and acting in the light of the interests of those other species. Even before genome sequencing, however, philosophers pointed to the relevance of qualities in nonhuman species such as sentience, which are arguably more salient than a genome sequence alone. Another issue concerns the self-perception of the human species in the light of the large overlap of genome sequence with other species. Rather than downgrading the perception of human specialness, however, another response to this is to argue that the sequence itself cannot be the place where human identity or distinctiveness is to be found (not withstanding the rhetoric of the common heritage of humanity and the iconic status of the germline). Rather, given the surprise about the smaller than expected number of genes identified by the HGP, the appropriate response may be that genes cannot explain as much about human life as had been previously thought.

Caveats may be recognised in relation to epigenetics and human futures as well. As regards epigenetics, in so far as there is acceptance that there are ethical and self-perception implications, the idea of a more environmentally aware human may be preferred to the idea of a more environmentally situated concept of the person.

Finally, as was mentioned earlier, when we turn to discussion of gene editing and the future, for some this is very clearly a matter of urgency for *human* survival, in part in the face of threats from pandemics such as COVID-19 or climate change. It could be argued that reflection on the human genome and the associated rhetoric of the common heritage of humanity can be leveraged to reinforce the importance of the idea of a global human community, in which the similarities between humans are far greater than the differences. While such a view appears to be anthropocentric it is not individual-focused and has important implications for social and international justice. What is clear, however, is that gen-ethics will continue to evolve, as does the genome itself.

Bibliography

Capps, Benjamin, Ruth Chadwick., et al. 2017. "Falling Giants and the Rise of Gene Editing: Ethics, Private Interests and the Public Good." *Human Genomics 11: 20.* https://doi.org/10.1186/s40246-017-0116-4.

Capps Benjamin, and Z. Lederman. 2015. "One Health and paradigms of public biobanking." *Journal of Medical Ethics* 41(3): 258–62.

Chadwick, Ruth. 1998. "The Communitarian Turn: Myth or Reality?" *Cambridge Quarterly of Healthcare Ethics* 20(4): 546–53.

———. 2017. "What's in a Name: Conceptions of Personalized Medicine and Their Ethical Implications." *Lato Sensu: revue de la société de philosophie des sciences* 4(2): 5–11.

Chadwick, Ruth., and Kåre. Berg. 2001. "Solidarity and Equity: New Ethical Frameworks for Genetic Databases." *Nature Reviews Genetics* 2(4): 318–21.

Chiapperino, Luca., and Giuseppe. Testa. 2017. "The Epigenomic Self in Personalized Medicine: Between Responsibility and Empowerment." *Sociological Review* 64(Suppl 1): 203–20.

Harris, John. 2016. "Germline Modification and the Burden of Human Existence." *Cambridge Quarterly of Healthcare Ethics* 25(1): 1–6.

Howard, Timothy. D., Fang-Chi Hsu et al. 2016. "Changes in DNA Methylation Over the Growing Season Differ Between North Carolina Farmworkers and Non-Farmworkers." *International Archives of Occupational and Environmental Health* 89(7): 1103–10.

Hughes, James, Nick Bostrum and Jonathan D, Moreno. 2007.., "Human vs Posthuman." *The Hastings Center Report* 37(5): 4–7.

Human Genome Organisation (HUGO). 1996. *Statement on the Principled Conduct of Genetic Research.* Available at http://www.hugo-international.org/hugo/conduct.htm.

———. 1998. *Statement on DNA Sampling.* Available at http://www.hugo-international.org/hugo/sampling.html.

———. 2000. *Statement on Benefit Sharing.* Available at http://www.hugo-international.org/hugo/benefit.html.

———. 2002. *Statement on Genomic Databases.* Available at http://www.hugo-international.org/hugo/HEC_Dec02.html.

Human Genome Organisation Ethics Committee. 2007. "HUGO Statement on Pharmacogenomics (PGx): Solidarity, Equity and Governance." *Life Sciences, Society and Policy* 3(44). https://doi.org.10.1186/1746-5354-3-1-44.

Lander, Eric S., Linton, L., Birren, B., et al. 2001. "Initial Sequencing and Analysis of the Human Genome." *Nature* 409: 860–921.

Le Page, M., "What your gut says about you." *New Scientist* 245(3265): 6.

Mulvihill, John J, Benjamin Capps., et al. 2017. "Ethical Issues of CRISPR Technology and Gene Editing through the Lens of Solidarity." *British Medical Bulletin* 122(1): 17–29.

Richardson, Sarah.S., Cynthia R. Daniels et al. 2014. "Don't Blame the Mothers." *Nature* 512: 131–32.

Savulescu, Julian., and Singer, Peter. 2019. "An Ethical Pathway for Gene Editing." *Bioethics* 33(2).

Sparrow, Robert. 2019. "Yesterday's Child: How Gene Editing for Enhancement Will Produce Obsolescence – and Why It Matters." *American Journal of Bioethics* 19(7): 6–15.

Tahira, Aisha., and Mark. Agius. 2012. "Epigenetics and Migration – Considerations Based on the Incidence of Psychosis in South Asians in Luton, England." *Psychiatr Danub* 24(Suppl 1): S194–S96.

UNESCO. 1997. *Universal Declaration of the Human Genome and Human Rights.* Paris: UNESCO.

Waterson, Robert H.., Eric S. Lander, and Richard K.. Wilson. 2005. "Initial Sequence of the Chimpanzee Genome and Comparison With the Human Genome." *Nature* 437: 69.

Part II

Bioethics and Posthumanism in Dialogue

4 Questioning the Politics of Human Enhancement Technologies

Thomas Hobson and Anna Roessing

Within the last decade, research in the biological sciences has revolutionised our understanding of life processes, biological entities, and the boundaries of the organic and inorganic world. Gene-editing technologies, particularly CRISPR, have accelerated the speed of gene editing while also dramatically reducing the cost. Frequently cited as "the most important innovation in the synthetic biology space in nearly 30 years" (Basulto 2015), its range of potential applications is apparently near-endless. Notably, proponents of gene-sequence editing have foregrounded its potentially transformative uses in agriculture, fuel-production, computing, and medicine (The Royal Society 2018).

In academic political and social studies, advanced technologies like CRISPR and other biotechnologies have generally occupied a rather peripheral location. There have, however, in recent years, been several contributions that engage critically with the ways that advanced and cutting-edge techno-sciences interact with humanity, society, and our ontological and normative preoccupations.[1] While it is still commonplace for technology and science to be treated as exogeneous and separate from the political world, many scholars advance the thesis that to live in modernity is to have an "inescapably technological existence" (McCarthy 2017). It is to this growing and invaluable body of work that our efforts here are intended to contribute. In this chapter, we utilise the approaches of Science and Technology Studies - understood as allied with the critical posthumanisms discussed elsewhere in the volume - to explore the political dimensions embedded in human enhancement technologies and to challenge the transhumanist assumptions underpinning prevailing technological imaginaries.

Occasionally, the focus of politicians, the media, and the general population is drawn more keenly towards the present reality and future possibilities of human augmentation research. Often this occurs when a major, and generally controversial, event occurs – billed as either breakthrough or catastrophe. The announcement in 2019 of the "CRISPR-babies" (see for instance Ledford

DOI: 10.4324/9781003020707-7

2017), for example, captured the imagination of people all around the globe. Serving as a powerful demonstration of the incredible potential of this technology, Chinese biophysics professor He Jianku told the world of how he had edited the genome of twins in order to endow them with HIV immunity.[2]

Press coverage, and the responses of experts, governance organisations and the international community, generally framed the announcement in one of two ways. Firstly, much of the criticism and concern centred on the uniquely *Chinese* character of this research – drawing attention to (arguably rather orientalist) tropes of secrecy, state control and disregard for ethics and regulation.[3] This line of criticism appeared to simultaneously imply that not only could this research *only have taken place* in an authoritarian and apparently lesser-moral state such as China, but also that, if it *had* somehow taken place in the United States or "the west," that the cultures of scientific openness and democratic science would have been less troubling.[4]

Secondly, many people, including researchers in the sciences, philosophy, and social and political sciences, focused their concern on how this research forced us to confront important moral and ethical questions. Rather than focusing on the ethical issues of the experiment itself, these critics instead asked how this type of human genome alteration casts uncertainty on our long-held and stable notions of human agency, subjectivity, and ontology. These critics appeared to suggest that, in altering the genome of foetal twins, Professor Jian and his laboratory had made it urgently necessary for us to ask questions about what exactly the human *is* and what the human *could be or ought to be* in the medium-term future (Krimsky 2019).

Alongside these branches of rather critical coverage and response was a third set of reactions to He's "CRISPR-Maybes." These responses were best-typified by those scientists, technologists, and investors who saw the experiment as a significant triumph for human augmentation research, and ultimately as a marker of significant progress in humanity's ability to edit, and to improve, itself. These transhumanist[5] visions are similarly concerned with troubling the boundaries of what it means to be human, and exactly what value such a categorisation can, or ought, to hold,[6] yet they generally frame these technological advances as the *solution* rather than the *problem* in reference to the moral and ethical quandaries of the human condition.

What is curious, and most certainly conspicuous by its absence, was a near total lack of discussion that consciously framed the world's first gene-edited babies as a political issue. That may initially seem to be a strange contention, given the exhaustive focus given to the character of science and technology – and particularly to those technologies that shape human behaviour or biology – in China. However, while it is true that there was a great deal of space devoted to conversations about China's authoritarian tendencies and the apparent lack of transparency in Chinese scientific culture, there was very little work undertaken on the substantive links between China's socio-political order, its norms and ideals, and the modes and meanings of scientific innovation it fostered. That is to say: while many people were very keen to locate their objection to – or fear of – these gene-edited human infants as a particularly *Chinese* problem, there was relatively little effort to trace or uncover the relationships between social and political orders and the types of techno-science that these orders strive towards.

Engaging with the tools and concepts of Science and Technology Studies researchers, most notably Shelia Jasanoff and her *Sociotechnical Imaginaries*, we endeavour to bring the *political* into the problematisation of human augmentation study and critique. This chapter does not attempt to provide either a historiographic or genealogical analysis of an event or global moment in advanced biotechnology and human augmentation. Rather, it aims to unpack the intimate relationship between technology and its political ecology that often evades the critical agenda of researchers, journalists, and regulators focusing on the ethical dimension of technological change.

Specifically, we outline what we understand as a mutually constitutive relationship between the cultural and political values, meanings, and mobilisations of purpose that steer technological developments. At the same time, we argue that technology is itself a powerful ordering instrument, with its design, distribution, and uptake conditioning how we perceive our world – its potentialities, its problems, and the means through which we can address them. Any investigation into technology and its central function in the organisation of societies, accordingly, requires broader attention to its thick contextual environments.

As part of this endeavour, we attend to the different dimensions at which the politics of technology come to matter. That is for one, as is outlined with the reference to the recent controversy surrounding the "CRISPR Babies," the close entanglement between geopolitics, foreign policy, and strategic interests in national technological leadership as a form of symbolic and material power. Second, we engage with the intimate relationship between technological visions of design and notions of what constitutes the modern state and its society. As such we link technology to political rationales emerging within the last decades seeking to manage uncertainty for increasing societal control. Lastly, we address the politics surrounding expertise and power in the scientific and technological project. In pointing to the plethora of interest groups emerging arguably outside established institutions, we reveal the political and moral economies deployed by the broad actor landscape competing over the visions and purposes of biotechnology and "biohacking."

Human Augmentation and Transhumanism

Research in synthetic biology and gene-editing has also held a long-standing and close association with human augmentation and the trans-humanist tradition. While undoubtedly aware of the controversy it has created, a number of prominent figures have openly advocated for the radical potentialities of human augmentation and editing through CRISPR or other genome-editing tools.

Driven by the desire to transcend the constraints of our human being, to "perfect" humanity or to overcome death, proponents of human augmentation and genetic editing argue that these technologies could transform and radically improve human life and health. Indeed, many of the research programmes in CRISPR suggest that they may well be on to something. In recent years, the innovation of CRISPR-based therapies has opened-up scientific imaginations and expanded medical possibilities – potential therapeutic applications include, to name a few, treatments for Retinitis Pigmentosa, Huntington's Disease, Sickle Cell Anaemia and cancer (The National Academies of Sciences, Engineering, and Medicine 2017). However, alongside these medical research programmes, CRISPR has also become central to those who envision not simply fixing what is arguably broken – a categorisation that is far from uncontroversial in its own right – but instead fundamentally altering the human status-quo.

These transhumanist thinkers see CRISPR, and other advanced biotechnologies, as making it not only possible but morally necessary to enhance even the most essential neurological and metabolic human processes. Advocates such as Zoltan Istvan have argued that:

> With this type of gene editing tech, we have a chance to wipe out hereditary diseases and conditions that plague humanity. And we could also modify the human being to be much stronger and functional than it is. CRISPR could be one of the most important scientific advancements of the 21st Century. We should embrace it.
>
> (Istvan, quoted in Pearlman 2015)

The concerns of those who are critical of such transformative CRISPR editing of the human genome are starkly observable in this framing. The "slippery slope" (George Church in Baker 2016) from therapeutic to augmentation applications appears to be very steep indeed if their proximity in rhetoric is anything to go by. George Church – a noted proponent of CRISPR – has repeatedly emphasised that "enhancement will creep in the door" and that "human enhancements will come after very serious diseases and they will be spread by somatic gene therapies" (ibid.) As to the desirability or potentially undesirable consequences of such human modification, comparisons to Huxley's dystopian vision of a genetically divided society in a *Brave New World*, and to more recent cinematic depictions in *Gattaca*, are readily and frequently drawn.

Francis Fukuyama observes that "while society is unlikely to fall suddenly under the spell of the transhumanist worldview," the incremental advances of actual scientific progress may facilitate our neglecting the potential moral cost of human augmentation and editing of the genome. "The first victim of transhumanism might be equality" (Fukayama 2009) he goes on to argue. In much the same vein, though with perhaps even more fervour, George Annas has advanced that:

> it almost seems inevitable that genetic engineering would move homo sapiens into two separable species: the standard-issue human beings would be seen by the new, genetically enhanced neo-humans as heathens who can properly be slaughtered and subjugated. It is this genocidal potential that makes species-altering genetic engineering a potential weapon of mass destruction and the unaccountable genetic engineer a potential bioterrorist.
>
> (Annas 2000)

While these critiques do a great deal to challenge the utopian potentialities of CRISPR and genome editing advanced by their proponents, they leave us with little else in terms of understanding why and how they may advance these visions of the future, by what means they may become manifest, and why they are significant. One useful means for uncovering these questions is to emphasise the *political* and ultimately *contingent* character of the histories, contemporary developments, and prophesied trajectories of genome modification, CRISPR and human augmentation. In doing this, it is possible to mount a sustained challenge to notions of rationality or necessity in techno-scientific visions of human futures, and to interrogate the highly situated ontological, normative, and material contexts of those who advance them.

Sociotechnical Imaginaries

Sheila Jasanoff has argued that

> the dream of escaping the burdens of the human condition is older than old. What gives the contemporary discourse on posthumanism, or transhumanism, its novelty is not simply talk of breaking the fetters of

mortality or the limits of human cognition. The newness, some would say, lies in blurring the boundaries between technology and the human.

(Jasanoff 2016)

In this telling, it is twenty-first-century advances in science and technology which mean that historic fantasies of human mastery over even the most incontrovertible realities of nature exist as attainable possibilities.[7] Real-enough, at least, to give rise to applied research, regulation, and controversy. While it is certainly true that our techno-scientific capacities now mean that some laboratories do possess the capability to alter, for example, the genome of a natal infant, it is also true that technological capacities – viewed in isolation – do not tell the full story about contemporary engagements with human augmentation techno-sciences. Turning again to Jasanoff, it is crucial to note that advances in our scientific and technological achievements have also altered our understanding of what being human means and our imaginations of "what lies beyond, or after, humanness as we know it" (Jasanoff 2016).

Life Scientists engaged in genetic engineering research, and those who sponsor and promote their work, can be seen as part of a biotechnological avantgarde. The entrepreneur, biohacker, and funder Ryan Bethencourt goes as far as to characterise biotechnologists and biohackers in pseudo-military terms, as the "foot soldiers of the next revolution in biotechnology and medicine, willing to do what others can't or won't."[8]

Whether by intent or as a by-product of their research, these actors are engaged in work that continually tests, and sometimes supersedes, boundaries – both those set by biology and those that exist as socially defined ethical frameworks.[9] The laboratory, however, does not exist independently of the world that surrounds it.[10] Scientists, students, private and state sponsors, regulators, and academics are all enmeshed and embedded in the social and political contexts, norms, and preoccupations of the contemporary world. The technologies they work to create are similarly shaped by the landscapes and limitations of our collective imaginations.

Jasanoff and Kim (2009) argue that technoscientific projects are always embedded within powerful visions of social and biological life that are attainable through technoscience. They define sociotechnical imaginaries as:

collectively held, institutionally stabilized, and publicly performed visions of desirable futures, animated by shared understandings of forms of social life and social order attainable through, and supportive of, advances in science and technology.

(Janasoff 2015)

Technologies then articulate normative visions of what life is and how it should be lived. They are, in effect, always already embedded in historically and culturally situated sociotechnical imaginaries (Jasanoff and Kim 2009, 2015). That means, necessarily, that these visions both emerge from and embedded within the "ethical presuppositions . . . social and cultural blindnesses as well as . . . acknowledged hopes and fears" of a given context and time.

We argue that the task of uncovering these contexts, and of tracing the formation, contestation, and influence of these sociotechnical imaginaries, is particularly urgent when considering mobilisations of techno-sciences that so consciously seek to alter the boundaries and meanings of human life. In their study of posthuman imaginaries, Tirosh-Samuelson and Hurlbut (2016, 4) analyse how technoscientific projects serve as crucial sites of moral and cultural production, wherein the past and present are (re)viewed and (re)produced through the prism of contemporary techno-scientific common sense, and possible technological futures are then collectively envisioned. When these visions concern the future of humanity – in a social as well as biological sense – in the immediate, mid-term, and long term, then the political and material contexts that shape their normative content must be interrogated.

Designed Life and the Miracle of Technology

It is our understanding that a functionalist reading of genome editing and synthetic biology as a mere means to alter biological processes omits the fundamental levels on which these technologies themselves coproduce the purpose of technology – as well as its objects. For the Life Sciences, and disciplines such as synthetic biology and genetic engineering, that means they not only provide new ways to alter life processes and generate knowledge but they also crucially shape the very ontologies we have of life.

A similar issue is discussed in Ana Delgado's (2016) analysis of the imaginaries promoted by synthetic biology, specifically the idea centring on design. Therein life is categorised as "an engineering material" and amorphous substance that becomes exploitable through engineering technologies. A "designed life," contrary to notions of biology's messiness and constant modes of transformation, invokes a notion of stasis that allows a fixed design to be instrumentalised and technologically exploited. Rather than processes of biological mimicry, it provides an "alternative nature" with reduced complexity or "messiness."

While the ontological status of nature and human life seem to be foremost an ethical and philosophical question, the invoked notion of "ontological indifference" that accompanies technical innovation (as described by Peter Galison 2006) yields equally problematic political consequences. It assumes technological neutrality and political inertness that effectively hide the practices of powerful actors in the construction and dissemination of technology locally and globally. First, the recognition of power and hierarchy corrects technology's often claimed determinist effects on society in acknowledging "not that technology develops outside of human agency, but that it develops outside of *some* humans' agencies" (McCarthy 2013, 476, emphasis kept).

When biological design seeks to "overcome the inherent arbitrariness and irrationality of biology" it stands to reason that we should question the motives and intentions – both conscious and implicit – of the individuals and institutions that will do the designing. This analysis comprises both the practices and epistemological commitments that create the cultural-political ecology in which these new visions and purposes of "biohacking" gain hold.

The cultural critic Jairus Grove (2019, 62) sees modern sociotechnical modalities embedded within political rationales to manage present complexity and future uncertainties. Translated to scientific practices, modes of scientific positivism emerge that fetishise "discreetness" as a desirable outcome of analysis. As technological and scientific capacities render more of the universe observable, in more detail, than ever before, the present is often increasingly understood computationally through measurable units of analysis. In much the same vein, complexity in the future is increasingly seen as best addressed through enhanced capabilities, rather than through processes of prediction and pre-emption.[11] Bishop and Philips (2010) argue that the "miracle of technology" relates scientific and technologically mediated processes of abstraction and datafication to social norms of modernity that allow this technological episteme to remain unscrutinised. In effect, technological visions and modes of modernity form a self-referential relationship in which technology functions as "a synecdoche of the modern era's representation of itself as supporting and supported by an increased functionality and instrumentality" (Bishop and Phillips 2010, 6) and thereby becomes authoritative.

The potential of advanced biotechnology is often conditioned by our capacity to put the ideas of the discreetness of design into practice. However, the sculpting of a system according to an ideal model risks hiding or even reinforcing the problems and inequalities of actually existing social orders. That is to say, the biopolitics and epistemes that structure our *Lebenswelt* determine the conditions of illness and disease from the norm(al) and healthy, find reinforcement through the technological means that identify which "broken" systems can be fixed, through what means, and to what end.[12]

In the twenty-first century, the biological sciences experienced a reorientation to the disciplinary logics of engineering and computation. In line with larger disciplinary convergences, the life sciences saw the entrance of new actors, skillsets, and scientific paradigms (Keasling 2008). In making design and engineering key to research practice and purpose (Delgado 2016), the type of performativity that design as research practice aims at affording, the materiality it assumes and produces and the type of engineering practices it deploys, all iteratively add to the invisibility of the technological rationales of design at work. Consequently, hiding behind the neatness of engineered productions of nature, the imaginary of synthetic biology is functioning as a projection space for the visions of a wide plethora of actors that adapt the rationale of synbio to achieve their sociotechnical futures.

The many transhumanist visions of biohacking deployed to achieve a carefree future existence therefore, not only touch upon ethical questions regarding whether to augment human biology but also crucially on who these visions include. Principles of design and engineering therein appear as a continuum of sociotechnical instrumentality that even takes hold of the messy realm of biology. Promising human capability that will "match up" with machinic superiority, the figure of a perfected human lurks in the imagination of some self-styled visionaries advocating for the radical potentialities of human augmentation through genome-editing tools.

In uncovering what is contingent, subjective and context-specific in discourses and practices of biotechnological research that are more generally characterised as objective and functional, it becomes clear that much of what is seen as *attractive* in these transhumanist visions of perfectible humanity is also at the core of what is deeply worrying about them. Efforts to re-invent the rules and meanings of human life through technologies shaped by visions of machinic superiority appear incapable of, or disinterested in, the kinds of self-examination required to avoid the very real potentiality for these techno-sciences to reinforce and even accelerate the inequalities and prejudices that underwrote their emergence. Jasanoff has argued that

> Scientific knowledge, in particular, is not a transcendent mirror of reality. It both embeds and is embedded in social practices, identities, norms, conventions, discourses, instruments and institutions – in short, in all the building blocks of what we term the social. The same can be said even more forcefully of technology.
>
> (Janasoff 2004, 2–3)

The often-unconscious imaginaries that guide the means and ends of biotechnological research into human augmentation must be subject to sustained and critical interrogation. Without this, it is not hard to hard to imagine how the scientific practice and attendant discourses of human enhancement could serve to both reproduce and accelerate inequality and injustice. There is a clear and present danger that *designed life* risks elevating certain categories of human life and proscribing others. If this is to be avoided, then urgent biopolitical and normative questions need to be actively engaged, rather than eschewed in favour of epistemologically discreet technology ones.

Conclusion

We have experienced an unquestionably significant transformation of social life through technology. Not because technology is situated outside of any social-political relationships but because it is embedded within them. This entanglement allowed technology to arrive as a powerful ordering instrument for those at the forefront of technological design, distribution, and uptake, conditioning how we perceive our world, its potentialities, and problems to be fixed.

In its many shapes and forms, technology has fundamentally transformed the material conditions of modern societies, but also, crucially, as Hurlbut and Tirosh-Samuelson (2016, 117) remark, "the imagination of power and progress in contemporary social life." Powerful technical imaginaries hence gain hold as the purpose and political endpoint of modern science – as "an authentic expression of the human essence" (Hurlbut and Tirosh-Samuelson 2016, 3).

As much as material realities, the stories we tell about technology and technology-mediated futures matter. They shape the worlds we see as possible and the means through which we try to attain them. Current narrations suggest the inevitability of technology and technology-induced progress. Equally, our terms of reference and debate are sometimes limited by the constraints of our shared comprehension. We become trapped in sequences of tokenised frameworks comparing technical risks yet neglecting substantive discussion as to whether any alternative means are available. Andy Stirling, Cian O'Donovan and Beccky Ayre (2018) recently critiqued the hidden technocratic governance beyond this line of reasoning:

> It is striking how political imaginations are constrained by conventional debates. In areas such as energy, chemicals or biotechnology, for example, the choices for innovation are restricted to the balancing of "risk" and "benefit" in some singular and supposedly inevitable direction for advance. Too often, the issues are reduced simplistically to a spectrum from "forging ahead" to "falling behind", as if the direction were predetermined or self-evident. Aims at fair distribution are restricted more to hopes of "trickle down", than the characters of the innovations themselves.
>
> (Stirling, O'Donovan, and Ayre 2018)

As these visions perform and enforce moral understandings of *how* we ought to coexist, *whom* these visions include and by what (technoscientific) means these visions are sought to be realised, they yield both material and symbolic power. At the same time, these processes of power and domination allow the stabilisation of some visions over others. As a function of a *Zeitgeist* – such as contemporary cultural sensibilities and dominating paradigms, scientific innovation practices and ethics are historically and materially contingent. Also, as the product of hierarchies within and among societies, experts and expertise as much as the decisions of governments and bureaucracies condition the discovery and design of technical objects and further direct vast state resources into their uptake and social entrenchment.

Notes

1. A by no-means exhaustive list here might include work from a broad range of scholars contained in the edited volume: J. Savulescu, R. ter Meulen, and G. Kahane, eds., *Enhancing Human Capacities* [Online] (Wiley-Blackwell, 2011). Also see: H. Tirosh-Samuelson and J. B. Hurlbut, *Perfecting Human Futures* (Weisbaden, Springer, 2016), 1–32; and those within Daniel McCarthy's edited volume: D. R. McCarthy, *Technology and World Politics: An introduction* (New York, Routledge, 2017). The works of STS scholars such as Jasanoff and Kim, including contributions to the volume J. Jasanoff and S.-H. Kim, *Dreamscapes of Modernity* (Chicago, University of Chicago Press, 2015); the now touchstone text from R. Braidotti and M. Hlavajova, eds., *Posthuman Glossary* (London, Bloomsbury Publishing, 2018); and the more demonstrable attention paid by political scholars more generally, such as: E. Schwarz, *Death machines* (Manchester, Manchester University Press, 2018); D. Malet, *Biotechnology and International Security* (Maryland, Rowman & Littlefield, 2016).
2. The timeline of leaked and announced information regarding the "CRISPR-babies" Lulu and Nana can be traced in the following news stories: LePage, Michael. 2018. "CRISPR babies: more details on the experiment that shocked the world," Nature. Sourced at: www.newscientist.com/article/2186911-crispr-babies-more-details-on-the-experiment-that-shocked-the-world. And here: Ramsey, Lydia. 2019. "The Chinese scientist who claims to have edited babies' DNA has been sentenced to 3 years in prison. Here's a timeline of the controversy." Business Insider. Sourced at: www.businessinsider.com/timeline-chinese-scientist-claims-crispr-babies-2019
3. See, for example, coverage of China's biotechnology labs from the strategic and national security scholars such as: K. B. Kania and W. VornDick, *China's Military Biotech Frontier: CRISPR, Military-Civil Fusion, and the New Revolution in Military Affairs* (Center for New American Security, 2019), www.cnas. org/publications/commentary/chinas-military-biotech-frontier-crispr-military-civil-fusion-and-the-new-revolution-in-military-affairs; Or, this coverage from the New York times, which emphasises the alleged illegality of comparable genome research in the US and the West: G. Kolata, S.-L. Wee, and P. Belluck, "Chinese Scientist Claims to Use Crispr to Make First Genetically Edited Babies," *New York Times*, 2018, www.nytimes.com/2018/11/26/health/gene-editing-babies-china.
4. It is certainly worth noting that this common line of critique inadvertently performed many of the central imaginaries of the virtues and exceptionalism of American scientific culture. For further discussion of these, please see, for example: S. Cozzens and E. Woodhouse, "Science, Government, and the Politics of Knowledge," in *Handbook of Science and Technology Studies*, Revised Edition, eds. S. Jasanoff, G. E. Markle and J. C. Petersen (Thousand Oaks, CA: SAGE Publications, Inc., 1995), 533–553.
5. We follow here a definition of transhumanism that rests on Moore's overarching categorisation that:

> transhumanism is a class of philosophies of life that seek the continuation and acceleration of the evolution of intelligent life beyond its currently human form and human limitations by means of science and technology, guided by life-promoting principles and values.

> More, Vita-More, More, Max, and Vita-More, Natasha. The Transhumanist Reader Classical and Contemporary Essays on the Science, Technology, and Philosophy of the Human Future. Chichester, West Sussex, U.K.: Wiley-Blackwell, 2013.

6. These visions are centrally focused on transforming human life through technology. The techno-utopian character of them is typified, for example, in: J. Mercier, "Elon Musk and Transhumanism: Behind the Movement to Merge Humans With AI," 2019, Medium. Sourced at: medium. com/@jakemerci/jeffrey-epstein-elon-musk-and-transhumanism; In Zoltan Istvan's column at Vice Magazine, available at: www.vice.com/en_us/ topic/zoltan-istvan; and in multiple entries on Ray Kurzweil's personal blog (and elsewhere), such as Kurzweil 2015. "Transhumanist position on human germline genetic modification." Sourced at: www.kurzweilai.net/ transhumanist-position-on-human-germline-genetic-modification.

7. It is worth drawing attention here to the work of Donna Haraway on the cyborg, and the blurring techno-human boundaries through biotechnology and cybernetic advances, to which the present work is also indebted: D. J. Haraway, *Simians, Cyborgs, and Women: The reinvention of nature. Simians, Cyborgs, and Women: The Reinvention of Nature*, London, Taylor and Francis Group, 1991.

8. The invocation here of an irrefutable necessity and a bravery and determination to do what is required, even when an implied beneficiary group do not either understand the need or have the courage to address it, is arguably significant in the performance of shared imaginaries of purpose for biotechnologists – perhaps particularly those engaged in boundary work on morally complex projects such as human augmentation. This is a rhetorical positioning reminiscent of Lieutenant J.G. Daniel Kaffe, reminding Lieutenant Weinberg that "we live in a world that has walls, and those have to be guarded" in the climatic court room scene during *A Few Good Men*.

9. If one is dubious as to the accelerated pace at which CRISPR-based technologies are pushing these boundaries, a review of the ongoing challenges and regulatory evolution of the annual iGEM competition should provide sufficient evidence as to their scale. I gem safety.

10. For more discussion of the many ways the *world* makes its way into the *laboratory,* please see, for example: Bruno Latour and Steve Woolgar, *Laboratory Life* (Beverly Hills: Sage, 1979).

11. For a more thorough discussion of these issues, the authors recommend: Wenger, A., Jasper, U., & Cavelty, M.D. (Eds.). (2020). The Politics and Science of Prevision: Governing and Probing the Future (1st ed.). London, Routledge. https:// doi.org/10.4324/9781003022428

12. These trends are traceable through a sociotechnical imaginary emerging during the 20th century with the rise of network and communication technologies. Ideas such cybernetics allowed functionalist approaches of technology to increasingly govern societal processes. Developed in the mid-20th Century, in the wake of two World Wars, these approaches sought to predict enemy behaviour. After the end of the second world war, however, they lend themselves equally efficiently for means of socio-economic prediction and societal governance (Bousquet 2009).

Bibliography

Annas, George J. 2000. "The Man on the Moon, Immortality, and Other Millennial Myths: The Prospects and Perils of Human Genetic Engineering." *Emory Law Journal* 49, 753–782. Available at SSRN: https://ssrn.com/abstract=251850.

Baltimore, David et al. 2015. "Biotechnology. A prudent path forward for genomic engineering and germline gene modification." *Science* (New York, N.Y.) 348(6230): 36–8. doi:10.1126/science.aab1028.

Basulto, Dominic. 2015. "Everything You Need to Know About Why CRISPR Is Such a Hot Technology." *Washington Post*, November 4. Accessed December 5, 2015.

Bishop, Ryan and Phillips, John. 2010. *Modernist avant-garde aesthetics and contemporary military technology: technicities of perception.* Edinburgh, GB: Edinburgh University Press.

Bousquet, Antoine J. 2009. *The Scientific Way of Warfare: Order and Chaos on the Battlefields of Modernity.* London: Hurst and Company.

Braidotti, Rosi, and Maria Hlavajova, eds. 2018. *Posthuman Glossary.* London: Bloomsbury.

Delgado, Ana. 2016. "Assembling Desires: Synthetic Biology and the Wish to Act at a Distant Time." *Environment and Planning D: Society and Space* 34(5): 914–34.

Edwards, Paul N. 2010. *A Vast Machine. Computer Models, Climate Data, and the Politics of Global Warming.* Cambridge: MIT Press.

Endy, Drew. 2005. "Foundations for Engineering Biology." *Nature* 438: 449–53 www.nature.com/articles/nature04342.

Engel, Jodie, Janice Desir, and Jack M. Bernstein. 2014. *A Rose by Any Other Name: On Synthetic Biology, Genetic Engineering, and Societal Control of Technology.* www.synenergene.eu/brief.

Fukayama, Francis. 2009. "Transhumanism." *Foreign Policy.* foreignpolicy.com/2009/10/23/transhumanism.

Galison, P. 2006. The pyramid and the ring. Presentation at the conference of the Gesellschaft für Analytische Philosophie (GAP), Berlin.

Grove, Jairus. 2019. *Savage Ecology: War and Geopolitics at the End of the World.* Durham, NC: Duke University Press.

Gallegos, Jenna E., et al. 2018. "The Open Insulin Project: A Case Study for 'Biohacked' Medicines." *Trends in Biotechnology* 36(12): 1211–18.

Haraway, Donna. 1991. *Simians, Cyborgs, and Women.* London: Taylor & Francis Group.

Hurlbut, J. Benjamin. 2017. "A Science That Knows No Country: Pandemic Preparedness, Global Risk, Sovereign Science." *Big Data & Society* 4(2).

Hurlbut, J. Benjamin, and Hava Tirosh-Samuelson. 2016. *Perfecting Huma Futures: Transhuman Visions and Technological Imaginations.* Wiesbaden: Springer.

Jasanoff, Sheila, ed. 2004. *States of Knowledge: The Co-production of Science and Social Order.* London: Routledge.

———. 2016. "Perfecting the Human: Posthuman Imaginaries and Technologies of Reason." In *Perfecting Human Futures: Transhuman Visions an Technological Imaginations*, edited by Benjamin J. Hurlbut and Hava Tirosh-Samuelson, 73–95. Wiesbaden: Springer.

Jasanoff, Sheila, and Sang-Hyun Kim. 2009. "Containing the Atom: Sociotechnical Imaginaries and Nuclear Power in the United States and South Korea." *Minerva* 47(2): 119.

————, eds. 2015. *Dreamscapes of Modernity: Sociotechnical Imaginaries and the Fabrication of Power*. Chicago and London: University of Chicago Press.

Keasling J. D. 2008. "Synthetic biology for synthetic chemistry." *ACS Chem Biol* 3(1): 64–76.

Kera, Denise. 2014. "Innovation Regimes Based on Collaborative and Global Tinkering: Synthetic Biology and Nanotechnology in the Hackerspaces." *Technology in Society* 37: 28–37. www.sciencedirect.com/science/article/pii/S0160791X13000638.

Krimsky, Sheldon. 2019. "Breaking the Germline Barrier in a Moral Vacuum." *Accountability in Research* 26(6): 351–68. Web.

Kuhn, Thomas S. 1962. *The Structure of Scientific Revolutions*. Chicago and London: University of Chicago Press.

Kung, Stephanie H., Sean Lund, Abhishek Murarka, Derek McPhee, and Chris J. Paddon. 2018. "Approaches and Recent Developments for the Commercial Production of Semi-synthetic Artemisinin." *Frontiers in Plant Science* 9.

Ledford, Heidi. 2017. "CRISPR Fixes Disease Gene in Viable Human Embryos." *Nature News* 548(7665): 13.

McCarthy, Daniel R. 2013. "Technology and 'the International' or: How I Learned to Stop Worrying and Love Determinism." *Millennium* 41(3): 470–90.

McCarthy, D. R. 2017. *Technology and World Politics. In Technology and World Politics*. New York: Routledge. https://doi.org/10.4324/9781317353836

Meyer, Morgan. 2013. "Domesticating and Democratizing Science: A Geography of Do-It-Yourself Biology." *Journal of Material Culture* 18(2): 117–34.

The National Academies of Sciences, Engineering, and Medicine (NASEM). 2017. *Human Genome Editing. Science, Ethics, and Governance*. Washington, DC: The National Academies Press.

O'Malley, Maureen A. 2009. "Making Knowledge in Synthetic Biology: Design Meets Klunge." *Biological Theory* 4(4): 378–89.

Pearlman, Alex. 2014. "Geneticists Are Concerned Transhumanists Will Use CRISPR on Themselves" in *Vice Magazine*. Article published 3/12/2015.

Perello, Edward. 2018. "CRISPR Genome Editing. A Technical and Policy Primer." *Microsoft Word*, November 19, 2018. Perello_final_corrected.docx (gmu.edu).

The Royal Society. 2018. "The CRISPR-Revolution: Changing Life." Conference Report. Accessed November 9, 2019. https://royalsociety.org/~/media/events/2018/03/crispr-revolution-tof/TOF-crispr-revolution-report.pdf?la=en-GB.

Schwarz, Elke. 2018. *Death Machines: The Ethics of Violent Technologies*. Manchester: Manchester University Press.

Stirling, Andy, Cian O'Donovan, and Becky Ayre. 2018. "Which Way? Who Says? Why? Questions on the Multiple Directions of Social Progress." *Technology's Stories* 6(2). www.technologystories.org/which-way-who-says-why.

Vincent, Bernadette B. 2014. "The Nature of Being a Microbe." In *BioScience* 64(8): 745–46.

Wenger, Andreas, Ursula Jasper, and Myriam Dunn Cavelty, eds. 2017. *The Politics and Science of Prevision: Governing and Probing the Future*. London: Routledge.

5   Biohumanities

Stefan Herbrechter

One of David Eagleman's speculative tales in *Sum: Tales of the Afterlives* begins like this:

> There is no afterlife of us. Our bodies decompose upon death, and then the teeming floods of microbes living inside us move on to better places. . . . Our death is unnoteworthy and unobserved by the microbes, who merely redistribute onto different surfaces. So although we supposed ourselves to be the apex of evolution, we are merely the nutritional substrate.
>
> (Eagleman 2009, 54–55)

What the neuroscientist Eagleman thus describes, not without a certain irony, coincides with a shift in the life sciences towards "new microbiology" (Cossart 2018; see also endnote 4). As a result of this shift, the true hero of evolution is no longer the human but microbial life. It is a (biocentric) shift one might also call the microbial turn in biomedicine (cf. Herbrechter 2018). The postanthropocentric focus on life itself this shift or turn implies coincides with a "nonhuman turn" (Grusin 2015) in the (posthumanist, or post-) humanities. It therefore follows that, as Catherine Belling explains,

> When *bios* – life – is liberated from [traditional] *biology* – the already-enculturated science of living things – bioethics becomes an endeavour that, while still unquestionably human, is imbricated in the concerns and claims of a biosphere that both enables and is threatened by human activity.
>
> (Belling 2016, 3)

And it is in this (imbricated bioethical) sense that "the humanities not only comment on the significance or implications of biological knowledge but

DOI: 10.4324/9781003020707-8

add to our understanding of biology itself" (Stotz and Griffiths 2008, 37), which, as I would argue, calls for the notion of "biohumanities."

Traditionally, bioethics is narrowly defined as "the discipline dealing with ethical issues relating to the practice of medicine and biology or arising from advances in these subjects" (*OED*). It is based on moral discernment relating to medical policy and practice and arising from the connections between the life sciences, biotechnology, medicine, politics, law, and philosophy. Joanna Zylinska's critique of traditional bioethics shows that, from a posthumanist point of view, bioethics needs to be extended both as far as the "bios" and the "ethical" is concerned. The three accusations that Zylinska levels at traditional (i.e. humanist and anthropocentric) bioethics are that it relies on "predefined normativity, human subjectivity and universal applicability" (Zylinska 2009, 6). By way of illustration of how to begin to address these limitations, I will be referring to two images, both title pages taken from popular science magazines. The idea behind this selection is one that I first outlined about ten years ago in my critical analysis of posthumanism as an emerging discourse in the humanities (Herbrechter 2013) – a discourse that developed out of theoretical positions taken up during the second half of the twentieth century and that were already challenging certain core humanist values, namely poststructuralism, deconstruction and postmodernism as well as feminism and postcolonialism. After a succession of fundamental controversies (wars and turns) about the role of theory, language, culture, and science, a number of questions returned, in the course of the 1990s, which signalled a shift towards posthumanism, at least in some parts of the humanities. These questions were: What is technology? What is the human? and What is life? None of these questions, which inform the mentioned "nonhuman turn" (or what one might also call the current "life wars"), are particularly new but what makes them worth returning to is the new context and historical situation in which they have regained their urgency. All are strictly speaking metaphysical questions – asking about the essence or truth of something at a time when (western) metaphysics is experiencing a radical crisis of legitimation, especially, as far as its underlying liberal humanism, anthropocentrism, human exceptionalism, and eurocentrism are concerned.

Asking, once again and with more urgency, the question concerning technology in a time when technological change is accelerating and when technology is increasingly seen as autonomous, but also as more and more invasive and "originary" (cf. Stiegler 1998), coincides with a time when some people (i.e. transhumanists) think that "we" (i.e. humanity) have reached a turning point at which certain technologies are either threatening or promising to take over. Needless to say that this remains highly controversial, utterly contestable and resistible, but it would be dangerous to simply dismiss the desire that is behind transhumanist dreams of technological enhancement, life extension and strong artificial intelligence, or their related projects of re-engineering life and the planet (i.e. geo-constructivism; cf. Neyrat 2019).

The question of what is or what makes us human is also being asked again at a time when anthropogenic climate change is threatening the survival not just of our own species but of life on this planet in general. Climate catastrophism, too, is usually framed by a kind of bioethical question, namely: what is a good life or what is sustainable living? Or, indeed, in what way may humans have to change in order to become a life-affirming and a life-preserving species? This produces a curious dilemma, namely that a strong sense of human agency and subjectivity is needed precisely at a time when the challenging of anthropocentrism and of human exceptionalism is put forward as a remedy for environmental degradation and adverse human geological impact.

Last but by no means least, the question of life and the living – arguably the central question for any bioethics worthy of that name – has been rediscovered precisely at a time when life itself has become an indispensable commodity (i.e. biocapital, biocapitalism, and bioeconomics) used by biotechnology and capitalist biopolitics to release its enormous market potential thanks to a combination of genetics and informatics (cf. Rajan 2006; Rose 2007; Cooper 2008; Braidotti 2008; Clough and Willse 2011; Braidotti 2013).

This sketch of our current situation, between "the Fourth Industrial Revolution and the Sixth Extinction," as Rosi Braidotti put it (Braidotti 2019, 2), with its combined threats of posthumanising technologies, anthropogenic climate change and aggressive biocapitalism, calls for a complex argumentative stance that is able to articulate both an ethics that is not human-centred and a politics in which human agency and responsibility are affirmed. This is what *critical posthumanism* as a programme, in my view, stands for.

To demonstrate this double imperative, we can refer to two contrasting visual illustrations. "Better than human – why settle for what you were born with?"[1] The cover of the May 6, 2006, edition of the *New Scientist* interpellates or addresses its (human) readers in the form of a visual face-to-face anchored by an implied imperative. It positions a subject as a human who is to be persuaded to see her body (gendering the body-to-be-transformed as female is certainly important in this context) not as a given but as something to be enhanced, extended, perfected. The technological framing of the constructed human face – cf. the circuit board-like pattern fusing the background with the face, the DNA-shaped earrings, the chipped necklace – sets the bio-digital cybernetic scene of interfacing bodies and technology and calls on the human subject to buy into the plasticity of digital ontologies. All kinds of bioethical (but also biopolitical) questions around feasibility, distribution, and denaturalisation arise for a traditional bioethics that is called upon either to legitimate or to caution against the transhumanist desire to overcome one's biological human condition through dematerialisation by gradually getting rid of the "wetware," until one either becomes, or is superseded by, virtual AI. The associated story in the magazine, "The Incredibles," frames this constructed future scenario with what I called "science faction," the deliberate fusion of science fiction and science fact (Herbrechter 2013, 114ff.): it specifically links the enhancement scenario to the Pixar animation movie *The Incredibles* – a perfectly "normal" and "everyday" family of super heroes (Lawton 2006). The underlying ideology, however, remains an entirely unreconstructed liberal humanist one that tries to make people believe that through technology they can be what they want to be. The implication is that humans "naturally" desire perfection; they are the privileged species that cannot stand still and are godlike in their ability to self-transform, etc.

There is a significant shift in perspective between the *New Scientist* cover and the June 2012 cover of *Scientific American*.[2] This, I would argue, also reflects a shift within popular posthumanist discourse itself (even though overlaps continue to exist, of course). The first phase, before the shift, was very much dominated by a cybernetic and digital imaginary which was basically neo-futurist and saw technology as an evolutionary driving force to be harnessed by superintelligent and "denatured" posthumans, or indeed, "exo-humans" (comparable to those space explorers who go on missions looking for new resources and life on exoplanets, or who are desperately looking for a new lease of life for something that might just about still be called "human" but will nevertheless be located somewhere "outside" existing humanity).[3] But then comes the realisation: what about humans and nonhumans and the mess they have created and find themselves in *here and now*?

The second cover image with its title – "Your inner ecosystem: In your body, bacteria outnumber your own cells by 10 to 1. Who is in control?" – is arguably much more "ecological" in the sense that it opens up the prospect of an entirely other bioethics (and biopolitics), one that is much more (bio) literally based on *postanthropocentric* premises. One could describe the "bacterialised" human shape on the cover by invoking the title of one of the chapters in Dorion Sagan's book *Cosmic Apprentice* (2013): "The Human is More Than Human: Interspecies Communities and the New Facts of Life." That the human is more than human is here precisely *not* seen as a techno-enhancement, or as the prospective transcendence of our animal bodies, but, instead, the cover argues that already from an evolutionary and biological point of view, we have never been human (at least not in any humanist sense). This could thus serve as an example of a weird kind of "bioenhancement," without any of the usual triumphalism.

If, "on a cell-by-cell basis . . . you are only 10 per cent human . . . for the rest, you are microbial" (Judson 2009), what might this spell out for bioethics and biopolitics? Human entanglement with the microbial is seen by many posthumanists as a further blow to "our" humanist narcissism. "New feminist materialists" like Rosi Braidotti, Moira Gatens, Claire Colebrook, Stacy Alaimo, Karen Barad, Donna Haraway, Myra Hird, Vicki Kirby, Jane Bennett, and Elisabeth Wilson, who have been arguing for a new understanding of the relationship between humans, their bodies and their nonhuman environment insist on stressing the "messiness" of complex matter-realities and corpo-realities. The ethico-political aim that critical posthumanism shares with these new materialisms which often emerge from a strong (feminist) affinity to the materiality of difference, is to find more ecologically and socially just forms of inter- and "intra-action" (cf. Barad 2003). They do so by breaking down the idea of a strong autonomy between (human) self and (nonhuman) other and by highlighting the co-constitution of the world through "biological, climatic, economic, and political forces" (Alaimo 2010, 2; for the (bio)ethical implications of posthumanist "corpo-realities" see MacCormack 2012). At the same time, they also critically investigate the contemporary extension of global biopolitics into the realm of the microbial, because the microbial level of life that inhabits every human, nonhuman and environment forms at once a connection with an ancestral past and with a posthuman future of life on this planet.

As opposed to *trans*humanist escapism and technoutopian geo-constructivism, an acknowledgement of the interconnectedness between humans, animals, microbes and matter in general can be understood as a new form of "worlding": "thinking in terms of microbes keeps us thinking in terms of being in this world and accountable to it, rather than envisioning an escape from it" (Buell 2014, 82). An evolutionary view that focuses on the microbial and its role in creating and sustaining all life thus leads to the notion of the "inextricable connectedness of all creatures on the planet, the beings now alive and all the numberless ones that came before" (Margulis and Sagan 1986, 9).

The eco-bio-philosophical and ethical conclusion that Lynn Margulis and Dorion Sagan draw from this alternative narrative, consequently, is one that acknowledges and favours entanglement, cooperation, and networking:

We are part of an intricate network that comes from the original bacterial takeover of the earth. Our powers of intelligence and technology do not belong specifically to us but to all life. Since useful attributes are rarely discarded in evolution it is likely that our powers, derived from the microcosm, will endure in the microcosm. Intelligence and technology, incubated by humankind, are really the property of the microcosm. They may well survive our species in forms of the future that lie beyond our limited imaginations.

(Margulis and Sagan 1986, 22)

This does not only impose humility on humans as a species – in fact, it problematises the very category of species, individuality, and identity.[4] The "new (micro)biology" that has been gaining influence is thus based on symbiogenesis. It inevitably also leads to a new medicine and to the emergence of new fields of knowledge that integrate developments within the life sciences and the medical or biohumanities.[5] Dorion Sagan spells out the ethical and medical implications of being-multiple for the biohumanities: "If the body-brain is not single but the mixed result of multiple bacterial lineages, then health is less a matter of defending a unity than maintaining an ecology" (Sagan 2013, 173).

For microbiome studies, as part of this new biology of entanglement, being human can be literally described as "gut feeling":

With respect to most biological research projects, human beings are so well integrated with their microbiomes that the individuality of human beings is better conceived as a symbiotic entity. Insofar as biological research is concerned, to be human is to be multispecies.

(Hutter et al. 2015, 1)

The (medical, ethical, ecological, political, etc.) conclusion that may be drawn from this symbiotic state is that of a common fate. Rather than being individuals, humans form "a community of *Homo sapiens* and microbial symbionts" (Hutter et al. 2015, 2–3). The fallout of this biological problematisation of (human) identity, which more or less coincides with decades of similar tenets in cultural theory and philosophy (notably in poststructuralism and postmodernism, and now posthumanism and animal studies), points towards an increasing convergence between certain sectors within the life science and the biohumanities.

Under these circumstances, ecology not only is relevant to an environment that is somehow *outside* the (human) body, but it also applies to everybody as such, to every *interior*, that is, to "your inner ecosystem" (as the second image and its associated article by Jennifer Ackerman illustrate; cf. Ackerman 2012). So, how can a posthumanist ethics mindful of "our" microbial symbiotic eco-ontology turn, what can undoubtedly still be reclaimed as a very humanist *memento mori* moment, into something politically more progressive and affirmative? If, for a "biophilosophy of the twenty-first century," as Eugene Thacker contends, "life = multiplicity" (Thacker 2008), in which individual human and nonhuman animal bodies, as well as plants or also "things" more generally, are not (or at least not only) singular subjects but are indeed irreducibly entangled in their past, present, and future environments, a biohumanities approach would necessarily have to work with a notion of bioethics and biopolitics that reflects the flatness (as well as the difference) and the entanglement of the multiplicity of life. This shift can then open up possibilities for "care" in "more than human worlds" (cf. Puig de la Bellacasa 2017) – which constitutes the bioethical and biopolitical programme for the biohumanties.

To recapitulate: a *trans*humanist bioethics threatens to make our current ecological predicaments a lot worse by consolidating human exceptionalism and biophobia. A *critical post*humanist version of bioethics instead would need to balance both the technological and the biological claims towards life by reconciling the deep ecological aspect of symbiogenesis with an acknowledgement of the originary technicity inscribed in life processes – a development modern biotechnology has prepared by making the distinction between organic and nonorganic life, biological and technological evolution increasingly problematic. Acknowledging this double (biotechnological) claim and its current context is the only adequate way to affirm and to come to terms with the material "messiness" of "our" time.

Notes

1. See *The New Scientist*'s website, at www.newscientist.com/issue/2551/.
2. The cover can be viewed at https://www.scientificamerican.com/magazine/sa/2012/06-01/. It is based on a design entitled "Microbiome" by Brian Christie. See https://bryanchristiedesign.com/microbiome.
3. The SF movie *Interstellar* (dir. Christopher Nolan, 2014) starts from a scenario in which "our" planet is suffering from ecocide, which leaves humanity with only two choices: look for survival on exoplanets and thus reinvest in NASA and space travel, or to commit to planet Earth and try and reverse climate degradation. The binary choice the movie seems to put forward is: do we need new astronauts or new farmers? It's the astronauts who turn out to be the heroes in the end, of course. Timothy Morton discusses *Interstellar* extensively in his *Humankind: Solidarity with Nonhuman People* (2017, 145ff.) in these terms.
4. The major challenge that the "new (micro)biology" referred to here poses to traditional post-Darwinian models of evolution – a challenge that, in turn, problematizes the very notion of species – lies in the

> extent and promiscuity of lateral gene transfer and the difficulties this raises for defining a "tree" of life, the importance of symbiosis and cooperation, and the reinstatement of the group [or species; SH] as an important – perhaps the most important – unit of selection.
>
> (O'Malley and Dupré 2007, 777–78)

See also the full quotation by Thiago Hutter, which concludes with the statement: "Insofar as biological research is concerned, to be human is to be multispecies" (Hutter et al. 2015, 1).

5. The realisation of the biological and evolutionary "entanglement" promoted by the notion of symbiogenesis also implies an increasing erosion of the boundaries between human and nonhuman (i.e. veterinary) medicine (cf. Viney et al. 2015) and an increasing awareness of the connection between health and ecology in times of climate change (cf. movements like One Health, GeoHealth, EcoHealth etc.; cf. Wolf 2014; Horton and Lo 2015; Almada et al. 2017; Zwyert 2017). Posthumanist, or postanthropocentric, perspectives are also increasingly seen to be of value in moving towards a more inclusive and holistic notion of "public health" (cf. Rock et al. 2014; Rock 2017; Cohn and Lynch 2017; Friese and Nuyts 2017; Andrews and Duff 2019).

Bibliography

Ackerman, Jennifer. 2012. "Your Inner Ecosystem." *Scientific American* 306(6): 20–27.

Alaimo, Stacy. 2010, *Bodily Natures. Science, Environment, and the Material Self.* Bloomington: Indiana University Press.

Almada, Amalia A., et al. 2017. "A Case for Planetary Health/GeoHealth." *Geo-Health* 1: 75–78.

Andrews, Gavin J., and Cameron Duff. 2019. "Matter Beginning to Matter: On Posthumanist Understandings of the Vital Emergence of Health." *Social Science and Medicine* 226: 123–34.

Barad, Karen. 2003. "Posthumanist Performativity: Toward an Understanding of How Matter Comes to Matter." *Signs* 28(3): 801–31.

Belling, Catherine. 2016. "Introduction: From Bioethics and Humanities to Biohumanities?" *Literature and Medicine* 34(1): 1–6.

Braidotti, Rosi. 2008. "The Politics of Life as Bios/Zoe." In *Bits of Life: Feminism at the Intersections of Media, Bioscience, and Technology*, edited by Annek Smelik and Nina Lykke, 177–92. Seattle: University of Washington Press.

———. 2013. *The Posthuman*. Cambridge: Polity Press.

———. 2019. *Posthuman Knowledge*. Cambridge: Polity Press.

Buell, Denise Kimber. 2014. "The Microbes and Pneuma That I Am." In *Divinanimality: Animal Theory, Creaturely Theology*, edited by Stephen D. Moore. New York: Fordham University Press.

Clough, Patricia Ticineto, and Craig Willse, eds. 2011. *Beyond Biopolitics: Essays on the Governance of Life and Death*. Durham: Duke University Press.

Cohn, Simon, and Rebecca Lynch. 2017. "Posthuman Perspectives: Relevance for Global Public Health." *Critical Public Health* 27(3): 285–92.

Cooper, Melinda. 2008. *Life as Surplus: Biotechnology and Capitalism in the Neoliberal Era*. Seattle: Washington University Press.

Cossart, Pascale. 2018. *The New Microbiology: From Microbiomes to CRISPR.* London: Wiley.

Eagleman, David. 2009. *Sum: Tales From the Afterlives*. Edinburgh: Canongate.

Friese, Carrie, and Nathalie Nuyts. 2017. "Posthumanist Critique and Human Health: How Nonhumans (Could) Figure in Public Health Research." *Critical Public Health* 27(3): 303–13.

Grusin, Richard, ed. 2015. *The Nonhuman Turn.* Minneapolis: University of Minnesota Press.

Herbrechter, Stefan. 2013. *Posthumanism: A Critical Analysis* (orig. German ed. 2009). London: Bloomsbury.

———. 2018. "Microbes." In *The Edinburgh Companion to Animal Studies,* edited by Lynn Turner, Undine Selbach, and Ron Broglio, 354–66. Edinburgh: Edinburgh University Press.

Horton, Richard, and Selina Lo. 2015. "Planetary Health: A New Science for Exceptional Action." *The Lancet* 386(November 14): 1921–22.

Hutter, Thiago, et al. 2015. "Being Human Is a Gut Feeling." *Microbiome* 3(9): 1–4.

Judson, Olivia. 2009. "Microbes 'R' Us." *New York Times,* July 21. Accessed October 7, 2019. http://opinionator.blogs.nytimes.com/2009/07/21/microbes-r-us/?_r=0.

Lawton, Graham. 2006. "The Incredibles." *The New Scientist* 190(2551) (May 13): 32–38.

MacCormack, Patricia. 2012. *Posthuman Ethics: Embodiment and Cultural Theory.* Farnham: Ashgate.

Margulis, Lynn, and Dorion Sagan. 1986. *Microcosmos: Four Billion Years of Evolution From Our Microbial Ancestors.* New York: Summit Books.

Morton, Timothy. 2017. *Humankind: Solidarity With Nonhuman People.* London: Verso.

Neyrat, Frédéric. 2019. *The Unconstructable Earth: An Ecology of Separation.* Translated by Drew S. Burk (orig. French 2016). New York: Fordham University Press.

Nolan, Christopher. 2014. Interstellar. *Paramount Pictures.*

O'Malley, Maureen A., and John Dupré. 2007. "Towards a Philosophy of Microbiology." *Studies in History and Philosophy of Biological and Biomedical Sciences* 28: 775–79.

Puig de la Bellacasa, Maria. 2017. *Matters of Care: Speculative Ethics in More Than Human Worlds.* Minneapolis: University of Minnesota Press.

Rajan, Kaushik Sunder. 2006. *Biocapital: The Constitution of Postgenomic Life.* Durham: Duke University Press.

Rock, Melanie J. 2017. "Who or What Is 'the Public' in Critical Public Health? Reflections on Posthumanism and Anthropological Engagements With One Health." *Critical Public Health* 27(3): 314–24.

Rock, Melanie J., et al. 2014. "Toward Stronger Theory in Critical Public Health: Insights From Debates Surrounding Posthumanism." *Critical Public Health* 24(3): 337–48.

Rose, Nikolas. 2007. *The Politics of Life Itself: Biomedicine, Power, and Subjectivity in the Twenty-First Century.* Princeton: Princeton University Press.

Sagan, Dorion. 2013. *Cosmic Apprentice: Dispatches From the Edges of Science.* Minneapolis: University of Minnesota Press.

Stiegler, Bernard. 1998. *Technics and Time 1: The Fault of Epimetheus.* Translated by George Collins. Stanford: Stanford University Press.

Stotz, Karola, and Paul E. Griffiths. 2008. "Biohumanities: Rethinking the Relationship Between Biosciences, Philosophy and History of Science, and Society." *The Quarterly Review of Biology* 83(1): 37–45.

Thacker, Eugene. 2008. "Biophilosophy for the 21st Century (2005)." In *Critical Digital Studies: A Reader*, edited by Arthur and Marilouise Kroker, 132–42. Toronto: University of Toronto Press.

Viney, William, et al. 2015. "Critical Medical Humanities: Embracing Entanglement, Taking Risks." *Medical Humanities* 41: 2–7.

Wolf, Meike. 2015. "Is There Really Such a Thing as 'One Health'? Thinking About a More Than Human World From the Perspective of Cultural Anthropology." *Social Science and Medicine* 129: 5–11.

Zwyert, Katharine. 2017. "Human Health and Social-Ecological Systems Change: Rethinking Health in the Anthropocene." *The Anthropocene Review* 4(3): 216–38.

Zylinska, Joanna. 2009. *Bioethics in the Age of New Media*. Cambridge: MIT Press.

6 Autonomous

Bioethics and/as Intellectual Property

Megen de Bruin-Molé

When we think about bodies and who they belong to, our minds may not jump immediately to intellectual property law or to posthumanism. We may also jump *straight* to these topics – there are endless examples of science fiction texts that deal with bioethics and their politics. Analee Newitz's (2017) novel *Autonomous*, for instance, depicts an unambiguously capitalist future, in which everything is owned by someone, everyone can buy and sell their own bodies freely, and copyright police hunt down various kinds of property infringers: appropriators and remixers of texts, of bodies, of data. These remixers act both out of their own interests and in the interests of those who have become desperate enough to sell their own bodies into indenture, or who cannot afford the "branded," legal versions of the latest medicines. The world of *Autonomous* is framed as exploitative, but complex – it is not without its moments of liberation and empowerment. In the later part of the novel, a police robot named Paladin is given ownership of her own systems, and muses on the implications of being able to make informed and autonomous decisions about one's own body:

> *Using software she had installed in her own mind, the bot generated a new key to encrypt her memories. For the first time in her life, the process worked. Her memories were locked down, and the key that the Federation held in escrow would be useless. It would take centuries for even the most state-of-the-art machine to decrypt what she had seen and known for the months she'd been alive. At last, she knew what it felt like to own the totality of her experiences.*
>
> *A profound silence settled around the edges of her mind, more powerful than a defensive perimeter in battle. Nobody could find out what she was thinking, unless she allowed it. The key to autonomy, she realized, was more than root access on the programs that shaped her desires. It was a sense of privacy.*
>
> (Newitz 2017)

DOI: 10.4324/9781003020707-9

Here, Paladin's autonomy is determined by a physical component – she literally owns her body and can make changes to her code – but also a more abstract feeling of privacy or individuality – she alone is responsible for her thoughts and decisions.

As in *Autonomous*, in our world there are many parallels between the laws and assumptions governing our physical or intellectual property, and those governing our bodies. In a digital culture, the right to ownership of our bodies and ideas is often framed as positive and democratic, even though we are increasingly aware of the ways these kinds of ownership are exploited under global capitalism. Our fictions reflect this ambivalence. As in *Autonomous*, in contemporary medical ethics autonomy and privacy are also central concepts (see, for instance, Straehle 2016; Rathor, Azarisman Shah, and Hasmoni 2016; Goodman 2016; Aboujaoude 2019; Zwitter 2019). In general, we can assume that bodies (or bodies of work) are part of a moral and cultural domain as well as a physical one. If all bodies and texts were morally or culturally neutral, we would not have to specify different laws for children's bodies and adult bodies, for instance, or for old texts and new ones.

To this end, it is also important to note that both bioethics and intellectual property law have important ties to Romantic humanism. While bioethics is a relatively recent discipline, many of its aims and sympathies in relation to human and nonhuman suffering are linked to a Romantic shift in understandings of ethics, and of the roles governing and legal systems should take in ensuring they are practised (see Amato 2014, 75–76). This understanding is inherently futurist and speculative in nature. As Dawid W. de Villiers argues,

> If, in its turn to the classical world, Romantic humanism should seem to be closely aligned with the preoccupations of the Renaissance *humanista*, it should be kept in mind that ultimately its interests lay not with the elucidation of classical culture but with the ongoing unfolding of human potential.
>
> (Villiers 2015, 30)

Traditionally, then, both bioethics and intellectual property law place value on rational, independent thought, and the idea that one body or text is entirely distinct from another. This is particularly true in the Western world, where individualism, capitalism, and rationalism converge. If "I think, therefore I am" as René Descartes famously claimed, then my body is essentially an object owned by my mind. The products of my mind and body should be mine to own as well. Attacks against one's body or one's writings are often framed as acts of theft or property destruction. This is complicated by the fact that both are increasingly seen through the lens of neoliberal, capitalist metaphors of (self-)ownership and (self-)improvement. In practice, these assumptions are changing, and the reality of ownership is often a bit more complicated. Both bioethics and intellectual property law are grappling with the limits of Western individualism and humanism. As political scientist Anne Phillips suggests, "framing threats to bodily integrity as if these were acts of trespass on private property is not helpful," since it too often leads us to take a reductive and potentially harmful approach to bioethics (Phillips 2011, 728). From a literary studies perspective, I would argue that our entire conception of private intellectual property is increasingly unhelpful, as cases of textual appropriation and remix can illustrate. I am far from the minority in this perspective. In the rest of this chapter, I aim to illustrate how these kinds of discussions in bioethics and in the intellectual property ethics of the posthumanities can inform each other. The rights we have over our bodies and over our texts are not as unrelated as we may think.

A full breakdown of the wide-ranging parallels between how humans and nonhuman creations (media, artworks, policy, etc.) are theorised and discussed could fill many volumes. Instead, this chapter will highlight some of the key shared concerns between bioethics and textual ethics, emphasising the value of continued discussions between these domains. In particular, I will be looking at the concept of "autonomy" in medical ethics, as compared to the issue of "authority" or authorship in the post-digital humanities and the remix movement. The chapter will use these threads to expose the ways in which such discussions are already posthumanist (that is, moving beyond conventionally humanist ideas about autonomy and ownership), and to encourage further dialogue between various disciplines and sectors.

The Politics of (Self-)Authorship and (Self-)Ownership

The key assumption still commonly shared by bioethics and intellectual property law is that people (and texts) are self-contained and self-sufficient. Under this assumption, all the answers to questions of authority that one might have about an individual body (of work) can be found in that body. In literary studies, this takes the form of schools like New Criticism, which tends to look at a text as a "self-contained, self-referential aesthetic object" (Schryer 2011, 29) rather than one determined by its author or readers. In bioethics, this is perhaps best exemplified by questions of autonomy and informed consent in the field of medical ethics. The British Medical Association's website describes patient autonomy (and respect therefore) as "probably the single most talked-about principle or concept in medical ethics," and "a cornerstone of medical law" ("BMA – 2. Autonomy or Self Determination" 2018). But to what extent can a person or text be considered truly autonomous? Of course, from the perspective of the law, someone must be responsible for managing the content and well-being of the body/ text, and for overseeing how that content is used or managed by others. In the case of texts, which as inanimate objects do not possess any conscious-ness or rational capacity of their own, this responsibility generally falls to the author.

In the case of people, if an individual cannot demonstrate autonomy or give informed consent, a relative or medical professional may take on this role. Turning again to the British Medical Association's website, we learn that

> *Two conditions are ordinarily required before a decision can be regarded as autonomous. The individual has to have the relevant inter-nal capacities for self-government and has to be free from external constraints. In a medical context a decision is ordinarily regarded as autonomous where the individual has the capacity to make the relevant decision, has sufficient information to make the decision and does so voluntarily.*
>
> *("BMA – 2. Autonomy or Self Determination" 2018)*

In terms of judging "capacity" in "competent adults," issues and debates immediately arise. Should patients with mental illnesses or disorders be judged as "competent"? Decisions around competence often lie more with the doctor than they do with the patient: as in Newitz's novel, though we are all "self-owned" in theory, in practice some individuals are more autono-mous than others. As Phillips writes, "from John Locke onwards, claims to self-ownership have played their part in the elaboration of radical as well as conservative traditions" (Phillips 2013, 20). In the field of bioethics, the consensus is that

> the traditional understanding of law in relation to the body has been rendered largely obsolete by the growth of medical sciences that first made possible the sale of parts of bodies, then the sale of reproductive material, and eventually the sale of genetic maps.
>
> (Richardson and Turner 2002, 40)

Likewise, the "globalisation of medical or body markets has meant that there is now little regulation of the use and purchase of the body" (2002, 40).

Similarly, in the world of texts, autonomy, authorship, and ownership are not always straightforward concepts. Copyright laws theoretically exist to protect everyone's autonomy and originality, but historically they do more to protect the interests of those already privileged or in power. Disney has more power to enforce its copyright, for instance, than a writer of fan fiction does. Intellectual property law is also part of a much larger history of structural discrimination. Writing specifically about gender, legal scholar Rebecca Tushnet argues that from

> the exclusion of intellectual property protection for cooking, fashion, and other traditionally feminine endeavors to concern over whether novels, or reading generally, were too female, women's subordinate status has transferred to their intellectual creations as easily as to their actual daughters.
>
> (Tushnet 2011, 2138)

Here and elsewhere, Tushnet draws useful parallels between textual reproduction and physical reproduction. Metaphors about textual appropriation are often directly about bodies and their ownership. In a 2014 article, Jenny Roth and Monica Flegel explore how writers of fan fiction have been painted by published authors not only as thieves of their copyright, but as kidnappers, slavers, and rapists. They write:

> *These iterations of the argument that fan fiction represents personal violation against one's family, as opposed to an intellectual property issue, often appear in professional authors' commentaries, suggesting that, at least for these authors, copyright as a concept in and of itself requires a more humanizing and personalizing narrative in order to describe the sense of wrong they feel when their characters are used in fan fiction. . . . Chris Byrne proclaims that fan fiction is "the literary equivalent of an obsessive stalker" (2007), while George R. R. Martin states, "My characters are my children, I have been heard to say. I don't want people making off with them, thank you" (2010). The rape/ stalking/kidnapping metaphor crops up as well in Gabaldon's posts, when she likens fan fiction to "someone selling your children into white slavery" (2010). Some fans also take up these analogies: one laments that slash fiction – fiction that foregrounds same-sex romantic pairings – is "rape, if not in deed then in intent."*
>
> *(Byrne 2007 and Martin 2010 in Roth and Flegel 2014, 903)*

There are also many instances where texts are framed as the children of their authors, and the principle of guardianship or stewardship evoked to justify why their theft is so problematic. Tushnet writes about the ways copyright owners "often fear remix because they fear the loss of control over their works (often conceived of – pun intended – as the kidnapping or molestation of the authors' 'children')" (Tushnet 2011, 2147). Like children, texts are often viewed as embodiments of a human legacy that will continue after the author/parent's death. Of course, from the perspective of many medical ethics policies, children themselves are not considered to be autonomous, and decisions about their care revert to the adults around them (see, for example, "Treating Children and Young People as a Medical Student – Ethics Toolkit for Medical Students – BMA" 2018). This does not take away from their moral right to be treated with care by other people. Does it (and should it) work differently for an author's texts?

Obviously there is a distinction, in most people's minds as well as in the law, between violating intellectual property and violating a person. But textual politics remains a useful tool through which to examine and to help us think about bioethics. Not least because texts themselves are important in the renegotiation and recategorisation of bodies, both historically and currently. In her book *Antebellum Posthuman* (2018), Cristin Ellis explores how the nineteenth-century writings of Frederick Douglass, Henry David Thoreau, and Walt Whitman "appropriate the materialist ontology, but not the racist politics, of antebellum racial science, producing an antislavery materialism that rebuts biological racism in its own empirical terms" (Ellis 2018, 2). And as highlighted by the aforementioned examples from fan fiction, in twenty-first-century digital culture, the question of who owns a text, and who can reproduce and remix texts, also has important implications for creators and collectors. Writing about vidding, or fannish productions that make use of proprietary songs and video, Tushnet suggests that remix is about "economic power: new technologies allow people with somewhat limited financial resources to talk back to mass culture in language that audiences are ready to hear" (Tushnet 2011, 2154).

Another basic question shared by textual ethics and bioethics is that of temporality. Do people and/or texts remain fundamentally the same despite the passage of time? In considering textual ethics, scholars are currently grappling with how to frame authorship and ownership in a post-digital culture, where the creative power of an author, a "once godlike power, to be 'the person who originates or gives existence to anything,' has democratized, to become everyone's responsibility, and tradable" (Hartley 2013, 43). If we are potentially all creators, and as creativity is increasingly revealed as a web of interconnected resources, privileges, and technologies rather than a mystical moment of genius, where do we assign ownership and responsibility? Jonathan Gray makes the case for studying "clusters of authorship" rather than individual authors (Gray 2013, 108), suggesting that

> [i]f we see texts as always in the process of *becoming* . . . we can not only reframe the traditional author's role in that process in a way that does not allow the figure undue power, but also be encouraged to look

for and examine the many other figures in the many other moments that author a text.

(Gray 2013, 89)

This is also a strategy that might apply to people, and benefit policymakers in bioethics.

Some policies already work along similar lines. Medical ethics generally offers provisions for people to refuse treatment for their future selves, for instance, but "they cannot commission treatment in advance" ("BMA – 7. Consent to Treatment – Adults Who Lack Capacity" 2018). The implication here is that the present self cannot foresee what technologies might become available, and is therefore not adequately informed about the best possible treatments for the future self. But this right to refuse treatment also holds for a "competent adult" who is later deemed to lack capacity – in all other aspects, at least in the eyes of the law, becoming a different self whose autonomy must be entrusted to another. Perhaps, in the end, a biopolitical model which takes into consideration the increasing interconnectedness of individuals, cultures, and systems at each point of a person's life would be more consistently ethical.

In the spirit of a posthumanistic approach to human ethics, perhaps texts and their related politics might become an entry point into some of these discussions. Critical posthumanism and the posthumanities are always focused around the figure of the human – how could they not be, when this is the perspective from which we are writing, teaching, and learning? However, the advantages of the posthumanities do not necessarily lie in their ability to interpret the nonhuman perspective, but rather in the way they use the very concept of a nonhuman perspective to approach and re-interpret the human from new angles.

Bibliography

Aboujaoude, Elias. 2019. "Protecting Privacy to Protect Mental Health: The New Ethical Imperative." *Journal of Medical Ethics* 45(9): 604–7. https://doi.org/10.1136/medethics-2018-105313.

Amato, Joseph A. 2014. "Suffering, and the Promise of a World Without Pain." In *Suffering and Bioethics*, edited by Ronald Michael Green and Nathan J. Palpant, 61–87. Oxford: Oxford University Press.

"BMA – 2. Autonomy or Self Determination." 2018. British Medical Association. December 6, 2018. www.bma.org.uk/advice/employment/ethics/medical-students-ethics-toolkit/2-autonomy-or-self-determination.

"BMA – 7. Consent to Treatment – Adults Who Lack Capacity." 2018. British Medical Association. December 6, 2018. www.bma.org.uk/advice/employment/ethics/medical-students-ethics-toolkit/7-consent-to-treatment-lacking-capacity.

Byrne, Chris. 13 December 2007. [Comment.] "Let's Get Transformative!" Whatever. Moderated by John Scalzi. http://www.whatever.scalzi.com/2007/12/13/lets-get-transformative/

Ellis, Cristin. 2018. *Antebellum Posthuman: Race and Materiality in the Mid-Nineteenth Century.* New York: Fordham University Press.

Goodman, Bryce. 2016. "What's Wrong With the Right to Genetic Privacy: Beyond Exceptionalism, Parochialism and Adventitious Ethics." In *The Ethics of Biomedical Big Data,* edited by Brent Daniel Mittelstadt and Luciano Floridi, 139–67. Law, Governance and Technology Series. Cham: Springer International Publishing. https://doi.org/10.1007/978-3-319-33525-4_7.

Gray, Jonathan. 2013. "When Is the Author?" In *A Companion to Media Authorship,* edited by Jonathan Gray and Derek Johnson, 88–111. Malden, MA: John Wiley & Sons.

Hartley, John. 2013. "Authorship and the Narrative of the Self." In *A Companion to Media Authorship,* edited by Jonathan Gray and Derek Johnson, 24–47. Malden, MA: John Wiley & Sons.

Martin, George R. R. 7 May 2010. "Someone is Angry on the Internet." Not a Blog. Moderated by George Martin. http://www.grrm.livejournal.com/151914.html

Newitz, Annalee. 2017. *Autonomous.* New York: Tor.

Phillips, Anne. 2011. "It's My Body and I'll Do What I Like With It: Bodies as Objects and Property." *Political Theory* 39(6): 724–48. https://doi.org/10.1177/0090591711419322.

———. 2013. *Our Bodies, Whose Property?* Princeton, NJ: Princeton University Press.

Rathor, M. Y., M. S. Azarisman Shah, and M. H. Hasmoni. 2016. "Is Autonomy a Universal Value of Human Existence? Scope of Autonomy in Medical Practice: A Comparative Study Between Western Medical Ethics and Islamic Medical Ethics." *IIUM Medical Journal Malaysia* 15(1). https://journals.iium.edu.my/kom/index.php/imjm/article/view/412.

Richardson, Eileen H., and Bryan S. Turner. 2002. "Bodies as Property: From Slavery to DVA Maps." In *Body Lore and Laws: Essays on Law and the Human Body,* edited by Andrew Bainham, Shelley Day Sclater, and Martin Richards, 29–42. London: Bloomsbury Publishing.

Roth, Jenny, and Monica Flegel. 2014. "It's Like Rape: Metaphorical Family Transgressions, Copyright Ownership and Fandom." *Continuum* 28(6): 901–13.

Schryer, Stephen. 2011. *The Republic of Letters: The New Criticism, Harvard Sociology, and the Idea of the University.* Columbia University Press. https://columbia.universitypressscholarship.com/view/10.7312/columbia/9780231157575.001.0001/upso-9780231157575-chapter-1.

Straehle, Christine. 2016. *Vulnerability, Autonomy, and Applied Ethics.* London, UK: Routledge.

"Treating Children and Young People as a Medical Student – Ethics Toolkit for Medical Students – BMA." 2018. The British Medical Association Is the Trade Union and Professional Body for Doctors in the UK. 2018. www.bma.org.uk/advice-and-support/ethics/medical-students/ethics-toolkit-for-medical-students/ethics-of-treating-children-and-young-people.

Tushnet, Rebecca. 2011. "Scary Monsters: Hybrids, Mashups, and Other Illegitimate Children." *Notre Dame Law Review* 86: 2133–56.

Villiers, Dawid W. de. 2015. "Catastrophic Turns: Romanticism, History and 'the Last Man'." *English Studies in Africa* 58(2): 26–40. https://doi.org/10.1080/00138398.2015.1083195.

Zwitter, Matjaž. 2019. "Autonomy and Its Limitations." In *Medical Ethics in Clinical Practice,* edited by Matjaž Zwitter, 35–44. Cham: Springer International Publishing. https://doi.org/10.1007/978-3-030-00719-5_6.

Part III
Exploring Posthuman Futures

7 A Posthumanist Critique of De-Extinction Science

Sarah Bezan

This chapter offers a deeply retrospective and prospective posthumanist critique of de-extinction science; one that examines the prospect of biotechnological interventions like de-extinction in the context of a paleolithic humanity that initiated the first extinctions of nonhuman animals, and which has since led into an Anthropocenic era in which accelerating species losses are considered potentially reversible. In what follows, I propose an answer to the question: *how can posthumanist theory inform an approach to the past 40,000 years of anthropogenic (or human-influenced) extinction, along with emerging biotechnological solutions that seek to reverse these species losses?* Although my primary aim is not to make a case either for or against de-extinction, I seek to show how human–animal relationships (and by extension, conceptions of species and of what it means to be human) are shaped by the retrospective histories and prospective futures of extinction processes. This analysis therefore takes seriously the concern posed by environmental ethicist Ben Minteer, who asks what de-extinction "might mean for us" (Minteer 2014, 261). Indeed, what de-extinction might mean for humans, and more importantly for animals, is the focus of this essay.

I present my argument in two parts. Through an analysis of "The Vitruvian Man of Mass Extinction" in the first part this chapter, I expose the limits of anthropocentric models that conceptualise extinction phenomena, and assess how the de-extinction debate is often embedded in a bombastic human exceptionalism that continues to limit the purview of ethical care of, and moral obligation to, critically endangered and extinct species. The second part of this chapter on "Retrospective and Prospective Posthumanism" advances a comparative analysis of "deep" extinctions (mammoths) and "shallow" extinctions (passenger pigeons) in order to demonstrate how de-extinction – as evidence of the desire to manage and engineer nonhuman life, death, and resurrection – in turn shapes the evolutionary timelines of both human and nonhuman animals. This critique leads into the final section

DOI: 10.4324/9781003020707-11

of the essay, which turns to a more speculative play on de-extincting *Homo sapiens* through an experimental line of inquiry that decentres the human from interpretations of the planetary past and future.

***

In accounting for the retrospective losses and prospective revivals of extinct species, I will begin by introducing a survey of arguments presented by conservation biologists, bioethicists, political thinkers, ecological managers, de-extinction scientists, and paleogenomics researchers from across the natural and social sciences. This survey broadly sketches the positive and negative outcomes of de-extinction science, the ethical relationships with de-extinct animal species (which extend to questions about animal welfare), and the web of connections that link de-extinction with other interventionist projects that propose to rewild and repopulate declining ecological habitats. Through a posthumanist and de-anthropocentric approach to this survey (outlined in further detail in each of the chapter's two sections), I explore the impact of de-extinction biotechnologies on extinct species and reflect upon the future of human–animal relationships in the context of the sixth mass extinction crisis.

As a relatively "new" technique in a centuries-old practice of conservation interventions (Novak 2018, 20),[1] de-extinction science is principally defined as the resurrection of an extinct species through the means of back-breeding, synthetic biology, and/or gene editing (see *https://reviverestore.org/passenger-pigeon-de-extinction/*). Given that de-extinction science generally creates hybrids rather than exact carbon copies of extinct species (Kaebnick 2017, 60), the prospect of biotechnological resurrection has recently begun to exert pressure on previously accepted species concepts within the scientific community. The idea of a species concept that can be categorised into phenetic, biological, ecological, and phylogenetic characteristics has, in the wake of biotechnology, become insufficient (Finkelman 2018, 11).[2] For de-extinction scientist Ben Novak, de-extinction demands a new category that he defines as *evolutionarily torpid species*: in short, species that are "falsely considered extinct" because they "persist in the form of cryopreserved tissues and cultured cells" and require biotechnological interventions in order to reproduce (Novak 2018, 2).

While theoretically sound, the introduction of this species concept has caused a number of conservationists to bristle with discomfort. Curt Meine, for instance, asserts that these kinds of interpretations of species are "narrowly reductionist, mechanistic, and technocratic" (Meine 2017, 12). A species, Meine asserts, is more than its genome: it is "the unique expression and evolution of its genome, through a vital population, interacting with and within a unique physical, biological, and social environment, over a unique period of time" (Meine 2017, 13). For other ecological managers and ethicists, concerns about who will physically care for early generations of de-extinct animal species (DEAS) influence what will come to serve as a good metric of "success" for de-extinction programmes (see Friese and Morris, 2; Browning, 795). Worries about the cumulative cryo-archivisation (Chrulew 2017, 284–85), patentability (Adams 2017, 538), and unmitigated capitalisation of extinct animals, accompanied by well-founded anxieties about the suffering of donors, surrogates, and offspring in the process of cloning extinct animals (Sandler 2013, 358),[3] likewise amplify the tenor of controversy over the use of such biotechnological tools. According to bioethicist Shlomo Cohen, de-extinction may have arguably "positive" outcomes (such as the advancement of technoscientific knowledge and the origination of "recreational value" of DEAS for nature tourism in sites like Pleistocene Park), but may also pose health risks to humans (through zoonotic disease) and potentially create hazards for other animals and ecosystems (Cohen 2014, 174).[4] For biologists Carrie Friese and Claire Morris, on the other hand, the outcomes of de-extinction science are tethered to what is a more foundational and critical question, namely: "*what kind of nature does de-extinction seek to make?*" (2). This inductive survey suggests that de-extinction has become more than a set of ethical dilemmas: it is a pressing provocation to reconsider what it means to be human in a world faced with an unprecedented rate of anthropogenic species losses.

A response to this provocation has perhaps never been more timely. Indeed, the present Anthropocenic era continues to spawn a number of interventionist projects – from climate engineering to rewilding and de-extinction programmes – that plan to safeguard environments from decline and obsoletion. But much like the term "Anthropocene" itself, it is the human (and human interests) that typically remain front and centre in these interventionist regimes. Schemes that tactically address declines in species biodiversity can often be deeply anthropocentric, structured by a set of human exceptionalist values that prioritise technological prowess and arbitrary preferences for "charismatic" species, which are in turn fundamentally influenced by an inordinate preoccupation with the centrality of human life in governing the future of human and nonhuman evolutionary timelines. As a counter to this, I probe the limits of anthropocentric models that fail to fully interrogate interventionist agendas. As I will show in the next section, thinking beyond human timescales and agencies facilitates a re-appraisal of the values that propagate the desire for species revival, and furthermore allows for a more capacious consideration of the interests of species that have been selected for de-extinction programmes.[5]

The Vitruvian Man of Mass Extinction

A critique of the anthropocentrism that is inherent to de-extinction science begins with an analysis of the conceptual frameworks that represent extinction phenomena. As one such example, paleontologist Ross Barnett's *The Missing Lynx: The Past and Future of Britain's Lost Mammals* (2019) analogises 580 million years of extinction through what Barnett calls a "bodily chronometer." To imagine mass extinction, Barnett writes, you must begin by stretching your arms out either side of your body. "Now," Barnett writes, "imagine that the distance from your left fingertip to your right fingertip is the history of all complex life on earth, starting with the first multicellular animal about 580 million years ago" (Barnett 2019, i). This bodily chronometer enables Barnett to orient readers to the major markers of mass extinction (and potential re-introduction of resurrected species) on a deep timeline that is mapped onto the static and homogeneous form of *Homo sapiens*. These markers include the first mass extinction of marine species (left elbow), the second mass extinction of ammonites (left shoulder), the third mass extinction known as "the Great Dying" (right shoulder), the fourth mass extinction in the Triassic and Jurassic periods (right elbow), and the fifth mass extinction of the dinosaurs (right wrist). The sixth extinction, Barnett explains, "starts at about where the nail on your right middle fingertip began growing this morning. Those micrometres of keratin stand for the last 50,000 years of death on earth" (Barnett 2019, i). *The Missing Lynx* further outlines the extent to which the human species has infiltrated every habitat on the earth (through to the ends of the human's very fingertips), thereby causing the loss of "mammoths and moa, dodos and diprotodonts, pampatheres and passenger pigeons, toxodonts and thylacines" (Barnett 2019, i). As Barnett insists, "the disappearance of so many species is not a coincidence. *We* are causing the sixth extinction" (Barnett 2019, i).

Barnett's bodily chronometer exemplifies what Tom Tyler has assessed in *Ciferae: A Bestiary in Five Fingers* as Greek philosopher Protagoras's claim that "man is the measure of all things." The epistemological anthropocentrism of this claim is Tyler's target, which he deciphers in his compelling treatment of *anthropos*, *Homo sapiens*, and the figure of "man," which is modelled by the human hand as a tool and a symbol of the human's presumed ontological and epistemological superiority over nonhuman life (Tyler 2012, 11). Extending Tyler's challenge to epistemological anthropocentrism, I read Barnett's bodily chronometer as a corollary of human exceptionalist imaginaries that similarly activates a broader claim about the human body as the container for worldly processes and the primary metric for the continuation and decimation of "life." While Barnett is certainly right about the role of the human species in initiating and perpetuating the sixth mass extinction crisis, I would like to more closely scrutinise the limits of conceptualising extinction phenomena through the form (and more specifically, the armspan) of the modern human. By figuring the human centrally in a 580 million year timeline, such an equation calls to mind the Vitruvian Man, which is for humanists the emblem *par excellence* of humanistic values including reason, ascendancy over nonhuman animals, and the achievement of bodily perfectibility (regarded as *telos* of human evolutionary progress).

In *The Posthuman*, Rosi Braidotti's view of the Vitruvian Man is that it is "an ideal of bodily perfection which, in keeping with the classical dictum *mens sana incorpore sano*, doubles up as a set of mental, discursive and spiritual values" (Braidotti 2013, 13). For Braidotti, the literal dismantling (by the post-1968 generation) and subsequent philosophical deconstruction of the Vitruvian Man laid the groundwork for a wave of feminist, anti-colonial, new materialist, and posthumanist thought. Yet the doctrine of the Vitruvian Man, which "combines the biological, discursive and moral expansion of human capabilities into an idea of teleologically ordained, rational progress," is a view of human capability that stubbornly persists today, albeit on an increasingly unstable footing (Tyler 2012, 13). The 2019 cover of *The Economist*, for instance, bears the image of the Vitruvian Man with his arms outstretched, enclosed in a planetary sphere constellated by a number of twenty-first-century ciphers and iconographic symbols. Among them are the four horsemen of the apocalypse, a tattoo of a double helix of DNA on the inner flesh of the Vitruvian Man's forearm, a stork carrying a new-born (designer) baby with an inscription of a barcode on the bag, and endangered animals, such as a Giant Panda and a keratin-scaled pangolin (now nearly brought to extinction due to poaching). Although the Vitruvian Man ostensibly represents the reach toward human achievement (seemingly always in balance and in proportion), *The Economist* cover belies a sense of approaching calamity and uncertainty about the use of genomic data and the exploitation of nonhuman animals. Indeed, we might read the cover of *The Economist* as a properly speculative – if not also precarious – appraisal of a world coming undone at our fingertips.

To view *The Economist*'s Vitruvian Man in this way is to interpret twenty-first-century Vitruvian doctrine as a polarised set of oppositions around environmental decline and progress. A case in point is Charles C. Mann's *The Wizard and the Prophet: Two Remarkable Scientists and Their Dueling Visions to Shape Tomorrow's World* (2018), which illustrates how scientists Norman Borlaug (the techno-optimistic wizard) and William Vogt (the prognosticating prophet) respond to global problems like the energy crisis and climate change. Mann's book foregrounds a version of the Vitruvian doctrine that is unsteadily poised between self-inflicted adversity and innovation. Likewise, proponents and critics of de-extinction science ostensibly fall into either category: those actively advocating for de-extinction (George Church, Stewart Brand, Ben Novak, and others) see technological and scientific methods as promising solutions to the sixth mass extinction crisis, while those raising concerns about the limits of these methods (including but not limited to Ben Minteer, Bruce Jennings, and Ronald Sandler) make a case for finely tuning and redirecting financial resources toward conservation programmes that will protect critically endangered species. While I risk oversimplifying the debate, my point is that arguments presented by both the wizards and prophets of de-extinction primarily rely upon an idea of the human as the evolutionary nexus of unceded power and promise. Disengaging from these kinds of deeply entrenched anthropocentric ideals requires a posthumanist approach that will reach far beyond the realm of human interests and in turn think more expansively about the meaning of extinction itself. To do so would be to dispute the validity of the Vitruvian Man reflected in Barnett's bodily chronometer, which represents the human as the organising principle around which extinction processes are made meaningful.

Such an approach would also consider, as Claire Colebrook argues in *Death of the Posthuman*, that "there was a time, and there will be a time, without humans," which "provides us with a challenge both to think beyond the world as it is for us, and yet mindful that the imagining of the inhuman world always proceeds from a positive human failure" (Colebrook 2014, 32–33). Certainly, recent fervent attempts to locate the origin of the Anthropocene (with the Pleistocene extinctions, the advent of agriculture, the onslaught of the Industrial Revolution, with the plutonium fallout of nuclear bomb testing, and so on) demonstrate a prurient and even androcentric compulsion to measure the length and duration of the human species in the deep time of planetary processes (Grusin 2016, iii). Colebrook, who explores how the human's (im)potentiality and self-annihilatory tendencies impact upon our moral and ethical obligations to nonhuman animals, offers a critical rejoinder to a human exceptionalist paradigm that magniloquently champions human ingenuity as the sole basis for the future of human-animal relations during the sixth mass extinction crisis (Colebrook 2014, 12). Understood in this way, the armspanning potential of *Homo sapiens* is measured only by its evolutionary contingency and fallibility. It is therefore only after an unreserved acknowledgement of the limits and potentials of de-extinction – and of its reliance upon the hubristic and prognosticatory ideals of Vitruvian doctrine – that it becomes possible to bring the interests of critically endangered and extinct species to the centre of the de-extinction debate.[6]

Retrospective and Prospective Posthumanism

Thinking more scopiously about the evolutionary entanglements of *Homo sapiens* with critically endangered and extinct species results in a more fulsome understanding of the meaning of anthropogenic extinction. This is because anthropogenic extinction operates on an expansive timescale of multiple geological eras (extending from the late Pleistocene epoch to the present) (Dawson 2016, 8). To consider the human in this way thus necessitates a more flexible and fluid definition of what it means to be human, along with a posthumanist interrogation of "what has been and what is" and what will be, over and against the idea of the human as "coming from a consistent idea of humanness and subjectivity" (MacCormack 2012, 8). This posthumanist approach functions according to a self-reflexive temporal twist that traces "a retrospective history where the past had already passed" (Dawson 2016, 6), and which observes the overlapping entanglements of the human *as* animal (Agamben 2004, 79). An approach that looks both backward and forward in its posthumanist critique works against stultifying conceptions of the human that regulate our responses to extinction phenomena (as with Barnett's bodily chronometer) while also extending our understanding of how the modification of nonhuman animal species has shaped, and continues to shape, the evolutionary pathways of *Homo sapiens*.

"Deep" and "shallow" extinctions, ranging from the woolly mammoth to the passenger pigeon (both candidates for de-extinction), serve as illuminating examples of the co-evolving human–animal relationship that extends beyond individual human lifespans.[7] The extinction of the *Mammuthus primigenius* during the last Ice Age was among a great number of anthropogenic extinctions that occurred over the course of 40,000 years (Greely 2017, 30). Likewise, the demise of the passenger pigeon, described in a wide range of written records during the nineteenth century in North America, starkly captures the role of human activity in decimating what were once robust populations of *Ectopistes migratorius*. While the mammoth constitutes a "deep" extinction (on a scale of thousands of years) compared with the "shallow" extinction of the passenger pigeon (on a scale of a hundred or more years), both examples showcase the role of humans in regulating nonhuman animal populations on a macroevolutionary scale.

It is this action – the poking of a proverbial finger into macroevolutionary processes – that exemplifies the humanist fantasy of omnipotent power over nonhuman life. But when read through the lens of a deeply retrospective and prospective posthumanism, these macroevolutionary histories and futures reveal what is an ambiguous and self-contradictory set of values that undermine the surmised ontological superiority of the human (if such a monolithic category as "the human" can be said to exist). These values, interpreted through the lens of what Gregory Kaebnick deems a "gardening ethic" that views "nature's value" as "derivative from the value of humans," indicates some sense of moral obligation to rectify the damage done through anthropogenic extinction (Kaebnick 2017, 63). However, I argue that these values more importantly testify to the function of *desire*. The question is not whether the concept of nature is static or dynamic; pristine or intermingled with humanity (Kaebnick 2017, 64): it is, rather, about the acutely antithetical desire to manage the vital capacities of life and living beings at a time when anthropogenic extinctions continue to mount at an ever-accelerating rate (see Rose 2006). Such a contradiction is the very hallmark of desire, which ebbs and flows over time and which is expressed unevenly by some (though not usually by all) human beings. The controversy surrounding the de-extinction of species like the woolly mammoth and passenger pigeon is therefore not just about ethics, but about a rapacious yearning for macroevolutionary authority: *a desire to make extinct, and a desire to de-extinct*. The truth is that if desire is a centrifugal force that spirals human action into opposing directions, then re-establishing meaningful connections with critically endangered and extinct species will have to begin not only with a challenge to Vitruvian ideals but also with a deeper reflection of the subjects and objects of desire that are circumscribed within extinction processes.

Conclusion: The De-Extinct Human

In the context of a comprehensive (and ever-growing) human evolutionary history, it is clear that there is no such thing as a pure human. Each member of the *Homo sapiens* species is a hybridised and deeply evolved form of some other latent and now-extinct hominid – from *Homo neanderthalis* to *Homo denisova* – that interbred tens of thousands of years ago. This hybridity challenges the idea that the human is external from, and therefore capable of exhibiting total mastery over, extinction processes. Constituted as a blend of now-extinct hominids, *Homo sapiens* is literally the embodiment of anthropogenic extinction. Death and desire, it seems, continue to lay the foundation for the retrospective and prospective making of the human.

The evolutionary history of *Homo sapiens* is critically important to understanding how anthropogenic extinction gives form and contour to human histories and futures. As Stephanie S. Turner suggests, the recent interest in species loss curiously

> coincides with the contested story of human evolution, and so it necessarily engages various explanations touching on a range of human

concerns – from where we are to where we have been and where we are going – that are at once biological, geopolitical, and historical.

(Turner 2017, 56)

Turner's analysis of extinction narratives and genome time indicates that the prospect for biotechnological solutions like de-extinction arises from a sense of nonhuman nostalgia or ritualised atonement (Turner 2017, 59). But as I suggested in the previous section, de-extinction is as much about resurrecting the fantasy of the human as it is about restoring lost species. Indeed, it was with much zeal (if not also some measure of alarm) that public audiences recently responded to the prospect of scientists growing Neanderthal brains in the laboratory (Devlin 2018). While fully cloning a Neanderthal is beyond the realm of possibility for now, such a fantasy of revival demonstrates how the lives of both human and nonhuman animals are subject to forms of retroactive and prospective re-creation (Hughes 2013). This fantasy incites a number of questions: *what would it mean to de-extinct the human?* Or, to rephrase the question posed by Friese and Morris: *what kind of human would de-extinction seek to make?* In a prospective future where the human has ceased to exist but has become, to use Novak's term, "evolutionarily torpid" – stocked away as cryo-banked remains – *what would it mean to resurrect us?* In order to bring these same questions to bear for nonhuman animals (and therefore to place their interests more centrally in the de-extinction debate), we must ask ourselves if the extinct and subsequently revived *Homo sapiens* would be human at all. As a speculative gambit that interprets expressions of desire as the guiding force of revivalist fantasies, this experimental line of inquiry allows us – if only for a moment – to decentre the human and its antithetical whims from an idea of the world before, after, and beyond *Homo sapiens*.

Notes

1. Novak makes a case for de-extinction as a conservation intervention that extends back to the 1830s (with the translocation of extant species to replace extinct animal populations). He writes that breeding and other conservation tools have served to manage nonhuman populations in sites around the world, thereby making the use of biotechnology a "new" technique in a centuries-old practice.
2. See also Robert et al., 1021–1031. Robert et al. raise doubts about the evolutionary benefits of de-extinction on local, global, and macro-evolutionary scales. They argue that the definition of "evolutionary proxies" has become meaningless given the unprecedented rate of species loss today.
3. For a consideration of the ethics of genetic engineering, see Cochrane 103–127.

4. Unlike Cohen, I doubt that the "recreational value" generated by extinct animals on exhibit would translate into increased public support for conservation programmes.
5. Candidates for de-extinction include the woolly mammoth, moa, auroch, great auk, Steller's sea cow, passenger pigeon, gastric brooding frog, and thylacine, among others. See Turner for a full table of proposed candidate species (408) or visit the Long Now foundation website at www.longnow.org.
6. While a discussion of the interests of whole species (versus individual animals) is beyond the scope of this chapter, it is worth noting that scientists like Ben Novak view de-extinction as an intervention that is in keeping with the interests of a species. However, in the case of the recovery of the passenger pigeon, the suffering of individual animals (including the suffering of laboratory-produced chicks of the hybrid passenger pigeon) is conspicuously absent in Novak's assessment of de-extinction efforts (see Novak, 20). Unlike Novak, I advocate for an approach that affirms the interests of individual animals and that also considers how (and in which ways) species might be said to have interests. Since the species concept is itself a discursive formation subject to cultural and political influences (see McKay, 253), I suggest that a critical examination of individual and species-level interests may aid in building a comprehensive assessment of de-extinction interventions.
7. Both of these projects are detailed on the Revive and Restore website. For more information on Ben Novak's passenger pigeon programme, visit https://reviverestore.org/events/tedxdeextinction/how-to-bring-passenger-pigeons-all-the-way-back/. Information about George Church, founder of the Woolly Mammoth Revival Project, can be found here: https://reviverestore.org/projects/woolly-mammoth/.

Bibliography

Adams, William M. 2017. "Geographies of Conservation I: De-extinction and Precision Conservation Progress." *Human Geography* 40(1): 534–45.
Agamben, Giorgio. 2004. *The Open: Man and Animal*. Translated by Daniel Heller-Roazen. Stanford: Stanford University Press.
Barnett, Ross. 2019. *The Missing Lynx: The Past and Future of Britain's Lost Mammals*. London: Bloomsbury.
Braidotti, Rosi. 2013. *The Posthuman*. Cambridge: Polity Press.
Browning, Heather. 2018. "Won't Somebody Please Think of the Mammoths? De-Extinction and Animal Welfare." *Journal of Agricultural and Environmental Ethics* 31: 785–803.
Cochrane, Alasdair. 2012. *Animal Rights Without Liberation: Applied Ethics and Human Obligations*. New York: Columbia University Press.
Cohen, Shlomo. 2014. "The Ethics of De-Extinction." *Nanoethics* 8: 165–78.
Colebrook, Claire. 2014. *Death of the Posthuman: Essays on Extinction, Vol. I*. London, UK: Open Humanities Press.
Chrulew, Matthew. 2017. "Freezing the Ark: The Cryopolitics of Endangered Species Preservation." In *Cryopolitics: Frozen Life in a Melting World*, edited by Joanna Radin and Emma Kowal. Cambridge, MA: MIT Press.

Dawson, Ashley. 2016. *Extinction: A Radical History.* New York: O/R Books.

Devlin, Hannah. 2018. "Scientists to Grow 'Mini-Brains' Using Neanderthal DNA." *The Guardian,* May 11. Accessed February 2, 2020. www.theguardian.com/ science/2018/may/11/scientists-to-grow-mini-brains-using-neanderthal-dna.

Finkelman, Leonard. 2018. "De-Extinction and the Conception of Species." *Biology & Philosophy* 33(32): 1–18.

Fuller, Errol. 2014. *The Passenger Pigeon.* Princeton, NJ: Princeton University Press.

Friese, Carrie, and Claire Morris. 2014. "Making De-Extinction Mundane?" *PLoS Biology* 12(3): 1–13.

Grusin, Richard. 2016. "Introduction: Anthropocene Feminism: An Experiment in Collaborative Theorizing." In *Anthropocene Feminism,* edited by Richard Grusin, i–xi. Minneapolis: University of Minnesota Press.

Hughes, Virginia. 2013. "Return of the Neanderthals." *National Geographic,* March 6. Accessed February 2, 2020. www.nationalgeographic.com/news/2013/ 3/130306-neanderthal-genome-extinction-cloning-hominid-science/.

Jennings, Bruce. 2017. "The Moral Imagination of De-extinction." *Recreating the Wild: De-Extinction, Technology, and the Ethics of Conservation,* special report, Hastings Center Report 47(4): 54–59.

Kaebnick, Gregory. 2017. "The Spectacular Garden: Where Might De-Extinction Lead?" *Hastings Center Report* (July–August): 60–64.

MacCormack, Patricia. 2012. *Posthuman Ethics: Embodiment and Cultural Theory.* London: Routledge.

McKay, Robert. 2018. "A Vegan Form of Life." In *Thinking Veganism in Literature and Culture: Towards a Vegan Theory,* edited by Emilia Quinn and Benjamin Westwood, 249–71. London: Palgrave Macmillan.

Mann, Charles C. 2018. *The Wizard and the Prophet: Two Remarkable Scientists and Their Dueling Visions to Shape Tomorrow's World.* New York, NY: Knopf.

Meine, Curt. 2017. "De-extinction and the Community of Being." *Recreating the Wild: De-Extinction, Technology, and the Ethics of Conservation,* special report, Hastings Center Report 47(4): 9–17.

Minteer, Ben. 2014. "Is it Right to Reverse Extinction?" *Nature* 509(May 15): 261–62.

Novak, Ben Jacob. 2018. "De-Extinction." *Genes* 9(548): 1–33.

Preston, Christopher J. 2017. "De-Extinction and Taking Control of Earth's 'Metabolism' *Recreating the Wild: De-Extinction, Technology, and the Ethics of Conservation.*" Special report. *Hastings Center Report* 47(4): 37–42.

Robert, Alexandre, C. Thevenin, K. Prince, F. Sarrazin, and J. Clavel. 2017. "The Ecology of De-Extinction: De-Extinction and Evolution." *Functional Ecology* 31: 1021–31.

Rose, Nikolas. 2006. *The Politics of Life Itself: Biomedicine, Politics, and Subjectivity and the Twenty-First Century.* Princeton: Princeton University Press.

Sandler, Ronald. 2013. "The Ethics of Reviving Long Extinct Species." *Conservation Biology* 28(2): 354–60.

Siipi, Helena, and Leonard Finkelman. 2017. "The Extinction and De-Extinction of Species." *Philosophy & Technology* 30: 427–41.

Turner, Derek D. 2017. "De-Extinction as Artificial Species Selection." *Philosophy & Technology* 30: 395–411.

Turner, Stephanie S. Spring. 2007. "Open-Ended Stories: Extinction Narratives in Genome Time." *Literature and Medicine* 26(1): 55–82.

Tyler, Tom. 2012. *Ciferae: A Bestiary in Five Fingers*. Minneapolis: The University of Minnesota Press.

"The World in 2019." *The Economist*. Accessed May 7, 2020. https://ukshop.economist. com/collections/the-world-in/products/the-world-in-2019.

8 Posthumanism and the Bioethics of Moral Responsibility

Matt Hayler

There's been an accident. A motorbike tried to squeeze between cars, over the speed limit, and caused a head-on collision. Or, rather, the rider did. It's an odd turn of phrase, "the bike/car/truck pulled out into traffic"; where are the people? In this accident, after all, we're not focused on the vehicles in the garage; it's the two people now in hospital that occupy our attention, the motorbike rider and the driver of the car he (and his bike) hit. And there are witnesses, everyone saw it; it's the rider's fault. He's responsible.

A humanistic model of responsibility necessarily places blame on individual human actors. Subjects, not objects, act (and act freely), and for the choices that they make, the price of freedom is responsibility:

> [t]he traditional humanist paradigm . . . views persons as free and responsible for their actions and ultimately for who they are and what they become as persons. As such, the self stands apart from the world in the most significant way, morally and metaphysically, originating acts spontaneously, freely, and independently. . . . [This] traditional concept of the core self . . . is the bedrock upon which our entire moral and legal system is based.
>
> (Hill 1997, 292–93)[1]

In this chapter, I want to argue that posthumanism, in its rejection of foundational humanistic principles, forces us to return to our models for justifying blame (and therefore praise) and to think again about who, or what, is responsible for actions in the world. Posthumanism's commitment to exploring humans' complex and constant relationship with their environments, alongside biological, psychological, and sociological discoveries which benefit from being understood within a posthumanistic frame, necessarily pry that door open, and this might offer grounds for new approaches to justice and the role of sanctions and reward. But the

DOI: 10.4324/9781003020707-12

challenges they present to our normative ideas of morality and justice are also troubling (to say the least) and need to be acknowledged and explicitly positioned within an ethical frame which accounts for their implications. It is striking how little engagement there has been with the implications of being entangled with the world for moral responsibility. In posthumanist research, humans are often positioned as porous, as entities through which other entities might run, and therefore as less in control than we tend to think we are and far from being the centre around which everything else orbits. For Francesca Ferrando, in her outlining of *Philosophical Posthumanism*, the human is best described as an "open network" which impacts on the planet in "broad and extensive, . . . entangled, subtle, and diffuse ways" (Ferrando 187). In *Posthuman Knowledge*, Rosi Braidotti describes the "knowing subject" as "not Man, or *Anthropos* alone, but a more complex assemblage that undoes the boundaries between inside and outside the self, by emphasizing processes and flows"; humans are "neither unitary, nor autonomous" but "embodied and embedded, relational and affective collaborative entities" (45–6). What links posthumanist thought, for Braidotti and Ferrando, is an agreement that "we currently need an enlarged, distributed, and transversal concept of what a subject is and of how it deploys its relational capacities" (Braidotti 2019, 40). But even as the seat of agency clearly changes in such approaches, little attention has been paid to the challenges posthumanism presents to our administering of blame and praise. Nancy Carranza's outline of agency in posthumanist thought is careful to point out that "a non-anthropocentric reconceptualisation of agency does not evacuate human responsibility" (2018), but her focus reflects the dominant trend in the field, that is, how human *systems'* environmental and discriminatory effects are rightful targets of anger and work for change. When it comes to how an individually entangled human might be differently held responsible for an event, however, very few posthumanists have touched on the implications of the field.

Jane Bennett joins other posthumanist thinkers in noting that agency is "a confederacy, and . . . human actants . . . themselves turn out to be confederations of tools, microbes, minerals, sounds, and other 'foreign' materialities" (36). But she also makes the rarer move of beginning to explore where this might lead, asking: "how would an understanding of agency as a confederation of human and nonhuman elements alter established notions of moral responsibility and political accountability?" (21). In seeing the human as a porous assemblage of human and nonhuman material, Bennett starts to realise that "this federation of actants is a creature that the concept of moral responsibility fits only loosely and to which the charge of blame will not quite stick" (28). She only expands on this a little further, but her conclusion gets to the heart of what I want to emphasise here:

> Autonomy and strong responsibility seem to me to be empirically false, and thus their invocation seems tinged with injustice. In emphasizing the ensemble nature of action and the interconnections between persons and things, a theory of vibrant matter presents individuals as simply incapable of bearing *full* responsibility for their effects.
>
> (Bennet 2010, 36; emphasis in original)

It is well beyond the scope of this chapter to either rigidly define posthumanism or to describe what a society which genuinely questioned individual responsibility would, could, or should look like. Instead, I want to suggest that any insight which troubles the boundaries of individual human agency necessarily contributes to reopening questions of moral responsibility, and that this is something which posthumanism must face up to. My approach here extends the work of Bruce Waller in *Against Moral Responsibility* (2011) with insights from contemporary science and posthumanism, and it requires that I give our rider a brain parasite, lead poisoning, an active microbiome, a badly designed bike, and a host of social influences and pressures, and to ask how blameworthy he really is.

Outlining Posthumanism and Moral Responsibility

For the purposes of this chapter, I take a broad view of posthumanism as any theory and/or practice rejecting the assumptions of humanism. And to establish a baseline for those assumptions, let's use Catherine Belsey's useful description of humanism ("a commitment to man, whose essence is freedom") and its subject ("the free, unconstrained author of meaning and action, the origin of history . . . unified, knowing, and autonomous") (8). By "unified," I mean definable and boundaried; by "knowing," I mean a rational actor which thinks in a recognisably and uniquely human way; and by "autonomous," I mean free and able to act alone. Any rejection of these features as descriptive of human subjects is, at least for my purposes here, posthumanistic.

Blame, however, is normatively based on humanist principles. When the rider of the motorbike caused the crash by their bad action, we see them as *morally responsible* precisely because we also see them as unified, knowing, and autonomous: they are a single agent, able to make a judgement, and acting of their own free will. They, and they alone, chose to act in a knowingly dangerous way and therefore deserve punishment for the consequences.[2] Driving a posthumanistic wedge in here might already seem unwise, not

least because the humanistic approach appears to capture something fundamental: humans typically "want to hold people morally responsible, and we
especially want to punish them for their wrongdoing – we feel a powerful
urge to strike back at them" (Waller, 94), and this "desire – indeed, the vis-
ceral biological need – to strike back at trouble can be found in chimpanzees
and rats. It existed long before humans" (49).

Throughout *Against Moral Responsibility*, Waller explores the predict-
able bleakness of this "strike back response" as grounds to unpick the ethics
of any system built upon it.[3] Similarly, if posthumanism aims at a better
understanding of human experience, and an improvement of our ethical
position, then this innate desire to punish needs to be accounted for. Waller
never mentions posthumanism, but in his focus on questioning the aims and
outcomes of allotting moral responsibility, he outlines a position which can
support, and benefit from, posthumanist theories of the human subject.

Waller's first step is to note that abandoning moral responsibility is not
equivalent to abandoning the judgement of good and bad acts: "[t]he question is not whether anyone ever does wrong, but whether those wrongdoers
justly deserve punishment" (Waller, 4). It's also not that actors cannot have
a good or bad character; "[n]o one is morally responsible for being bad or
behaving badly," Waller argues, "but this does not mean that no one has a
character with profound moral flaws" (5). Waller's case, instead, rests on
the poverty of evidence that bad acts, and bad characters, are the responsibility of the actors they emerge from or within.

Let's return to the bike rider. The act of driving recklessly is a bad act,
and maybe the rider has a bad character which makes him inconsiderate
of the ways in which his actions might affect others. What made him take
such a risk? If you, the honourable reader, had been driving the bike, then
you wouldn't have done the same thing – you wouldn't have undertaken the
bad act because you have a good character. You would have made a better
decision, and this puts you in the position of being able to judge the rider,
to blame them for their action, because you can imagine making the better
choice. But Waller wouldn't disagree that the rider did a morally bad thing
in risking the lives of other road users; instead, he would ask whether the
rider is *morally responsible for that bad action*. After all, if choice is
the important issue that separates you from the rider, did you choose to be the
kind of person who wouldn't act in a reckless way? Did the rider choose to
be reckless? Or are these accidents of character? Are you, in fact, *lucky* to
be capable of making better choices, and the rider *unlucky*?[4]

Waller frequently compares two actors in theoretical scenarios; let's call
them Anne and Betty. Anne, like you, makes good decisions; Betty, like the
rider, makes bad decisions. By simply blaming Betty and praising Anne,
however, we forget the histories of these women, of all the things which

came into play in making them who they are and the range of decisions that they are able to make. Anne, for instance, may have been lucky enough to have been born more intelligent in some useful way, have been better educated, and had better role models. Each of these aspects may have then shaped her cognitive abilities, limited her stresses, and expanded what she is now able to conceive of in comparison to Betty who was unlucky enough to have none of these gifts. It's not even that Anne may not have put work into being who she is, but the fact that she had the *capacity* to cultivate a good character capable of excellent decision-making is not solely down to her. As Waller puts it:

> if [Betty] had been a different person with different capacities and a different history, then she would have acted differently. If she had exactly the same history and resulting character as [Anne], she would have acted as [Anne] did . . ., but that fact has no relevance whatsoever for the question of whether [Betty] justly deserves blame or punishment in the world in which she actually lives.
>
> (Waller 2011, 27)

Here, then, is the essence of Waller's argument: who we are conditions what we do (and can conceive of doing) and is a product of historical luck[5] – if we accept that we are not morally responsible for the outcomes of luck then we are not morally responsible for our characters, and this demands, in turn, that we must reconsider our responsibility for our actions.

Further Challenges to Moral Responsibility

Daniel Dennett criticises Waller's position on the basis that society does its best to level out initial unfairnesses, and does so to such an extent that moral responsibility must return to the individual human actor:

> Life isn't fair. Some folks get a pretty raw deal through no fault of their own, and others thrive thanks to a great head start. . . . None of this is fair. . . . The state of nature isn't fair. That's why we have the institutions of civil society, to even the playing field *as best we can*, by minimizing the amplification of advantage and disadvantage that otherwise would probably occur. . . . You don't *have* to play the moral responsibility game, . . . [b]ut if you want to enjoy the benefits of living in a civilized society, you have to play. . . . [T]he fact that almost everybody agrees to play . . . means it is probably as close to a fair game as you could devise.
>
> (Dennett 2012; emphasis in original)

Dennett's critique, here, relies on two illusions: firstly, that societies work hard at levelling the playing field, and secondly that we choose to enter into moral responsibility as a contract in exchange for acceptance into society (a "choice" that Waller unpicks in his response to Dennett).[6] Unfairness is consistently drawn and redrawn across the same lines (race, class, gender, sexuality, disability, religion, etc.) and across generations. The impossibility of choosing whether to join such a system comes not only from exile being the only alternative, but also from our being born into hierarchies of unquestioned privilege and discrimination that have a tendency to occupy the horizons of our thinking, limiting our conceptions of how we, and the world, might be otherwise.

The continued faith in moral responsibility, despite the dramatic range of human luck, stems in large part from what Tom Clark calls "singling out the agent," that is, focusing on an individual ascribed the kind of humanistic subjectivity outlined earlier. As Clark describes it, singling out the agent focuses

> on particular characteristics . . . (e.g., being rational, reasons-responsive, sane, and uncoerced), while downplaying . . . causal history and the influence of the situation and systems of which [the agent] is inevitably a part. This selective emphasis . . . creates fertile psychological ground for generating attributions of agent-focused blame.
>
> (Clark 2012)

One of the primary characteristics of posthumanism, though, is its resistance to singling out the agent in precisely this sense.[7] I'm most influenced, here, by Donna Haraway whose *Staying with the Trouble* indicates how Waller's critique of moral responsibility might be assisted by posthumanistic thinking.[8] Throughout, Haraway describes human beings as "persons conceived as individuals" (217), and is continuously attentive to how "things and living beings can be inside and outside of human and nonhuman bodies, at different scales of time and space. All together the[se] players evoke, trigger, and call forth what – and who – exists" (16). This leads her to an unavoidable question: "[w]hat happens when human exceptionalism and the utilitarian individualism of classical political economics becomes unthinkable in the best sciences across the disciplines and interdisciplines? Seriously unthinkable: not available to think with" (57). For Haraway, the "what happens" is that humans need to reimagine human/animal hierarchies and social divisions in the face of the climate crises, habitat collapses, systemic discriminations, and species extinctions that industrialised, violent, and/or thoughtless entanglements create. But another thing that happens, I'm arguing, is an inherent re-opening of moral responsibility. In the following, I discuss a number of scientific enquiries which both support posthumanism's emphasis on entanglements and further challenge any theory of moral responsibility which ignores historical luck factors in its attempts to single out agents for blame.

Microbiomics, the study of the organisms that live on and within species (such as bacteria, fungi, and viruses), explores how human life, as all life, is riven through and wrapped up with other beings whose activity plays a role in both what we experience and who we are. In "How the Microbiome Challenges Our Concept of Self," Tobias Rees et al. discuss how

> the microbiome . . . plays a central role in the three processes that have traditionally been said to define the human self: the adaptive immune system of vertebrates that discriminates self from nonself . . . [;] the brain functions that underpin human personality and cognition[;] and the sequence of each person's genome that guides our unique phenotypic traits.
>
> (Rees et al. 2018, 2)

Rees et al. are careful to note that biology has long spoken of the ways in which nonhuman actors can *influence* these arenas of experience; the more profound move is from a model of influences to one of *co-construction*. As Gilbert et al. put it, "interactive relationships among species blur the boundaries of the organism and obscure the notion of essential identity[;] . . . organisms are anatomically, physiologically, developmentally, genetically, and immunologically multi-genomic and multispecies complexes" (326 and 331). The human-as-multispecies-complex (Haraway's unthinkable person as individual) manifests in a number of ways that relate to inherited luck, including gut microbes shaping behaviour and impacting on mental health (see Dinan et al.; Valles-Colomer et al.). Such ecological entanglements are far removed from our control, and whatever benefits or frustrations they bring can usefully be understood as additional aspects of the constitutive luck Waller describes.

For the rider in the hospital, perhaps the most significant microbial influence on his decision to pull into traffic could be the lesions left in his brain by a parasite. *Toxoplasmosis gondii* is a strange single-celled organism; it reproduces in the guts of cats, but from there it gets excreted in the cat's faeces. This, in turn, gets eaten by rodents and "Toxo's evolutionary challenge at that point is to figure out how to get rodents inside cats' stomachs" (Sapolsky). Rodents, of course, don't want to go near cats, or even to where cats have recently been; they have an innate aversion to feline pheromones. Robert Sapolsky, a neuroendocrinologist and Toxoplasmosis researcher, explains, however, that

> Toxo preferentially knows how to home in on . . . part of the [rodent] brain[, a] . . . region called the amygdala . . . which is all about pathways

of predator aversion. . . . This is a parasite that is unwiring this stuff in the critical part of the brain for fear and anxiety. . . . What you see is that the fear circuit doesn't activate normally, and instead . . . sexual arousal activates. . . . In other words, Toxo knows how to hijack the sexual reward pathway. . . . Somehow, this damn parasite knows how to make cat urine smell sexually arousing to rodents (Sapolsky).

(House et al. 2011)

This hijacking of rodent brains means that, rather than fleeing in fear, they head out *towards* the smell of cats, their heads full of parasites waiting to reproduce the moment that they're eaten, "Toxo's circle of life completed" (Sapolsky).

Humans can become infected with *T. gondii*, but the effects are typically minor. The human host might become a little ill, with "flu-like symptoms, such as: high temperature, aching muscles, tiredness, feeling sick, sore throat, swollen glands" ("Toxoplasmosis"), but, in the absence of underlying conditions, they tend to recover quickly, though the parasite always remains in a latent state. Over the last two decades, however, researchers have connected "latent" Toxoplasmosis infection with a series of changes in human behaviour including: i) depression, anxiety, schizophrenia, aggression, and suicide (see Flegr; Cook et al.; Xiao et al.; Ling et al.), ii) increased entrepreneurship (Johnson et al.), iii) excessive alcohol consumption (see Samojłowicz et al.), and iv) increased likelihood of traffic accidents (see Flegr et al.). With around 2 billion people worldwide estimated to have been infected, the implications of Toxoplamosis for understanding how, and how regularly, humans might be influenced by nonhuman life is significant; as Sapolksy notes,

no doubt [Toxo's] also a tip of the iceberg of God knows what other parasitic stuff is going on out there. Even in the larger sense, God knows what other unseen realms of biology make our behaviour far less autonomous than lots of folks would like to think.

(Sapolsky 2009)

Challenges to autonomy already appear in other areas of enquiry in biology, not least in the field of epigenetics.[9] An example that again demonstrates how constitutive luck might shape antisocial or risky behaviour is the link between exposure to lead, particularly in early (including pre-natal) life, and an increase in criminality, and particularly in violent crime (see Sampson and Winter; Wright et al.). These effects are particularly striking when looking at the effects of public health interventions: lead pipes installed in American cities have been connected to a 24% increase in murder rates in the early twentieth century, and the rise and removal of leaded fuel in cars seems to similarly be tied to a rise and decline in violent crime (see Feigenbaum and Muller; Mielke et al.) [air pollution more broadly has also been linked to violence and criminality (see Jackson et al.; O'Dell et al.)]. As Jessica Reyes notes in her own discussion of the effects of lead toxicity,

> social scientists struggle to explain trends in mental health and social behaviors, including learning disabilities, adolescent violence, teen pregnancy, and substance abuse. The complex variety of factors influencing these behaviors and conditions presents a challenge to researchers, and they are increasingly looking more closely at early life influences.
>
> (Reyes 2015, 1601)

Pollution can, of course, only be seen as a contributing, not determining, factor for elevated rates of crime, as noted by many of the studies cited above. Other contributing factors include the effects of poverty (see Webster and Kingston), poor nutrition (see Ramsbotham and Gesch), and levels and quality of education (see Lochner and Moretti), but no contemporary study suggests the increased presence of "bad" people in areas, or times, of higher crime. Intuitively, this would be a wholly insufficient explanation, but higher crime rates taken as an indicator of a higher concentration of morally worse people would also obscure a variety of factors, including how systemic racism and inequality manifest in practices of policing and governance alongside environmental exposures.[10] As Sampson and Winter argue,

> exposure to lead is unequally distributed, likely more so today than in the past when lead was "in the air" and thus almost everywhere. Contemporary lead exposure is linked, for example, to minority status and poverty at the individual level, as well as to racial segregation and concentrated poverty at the neighborhood level primarily because of the unequal distribution of dilapidated housing that contains remnants of lead paint.
>
> (Sampson and Winter 2018, 270)

To have to see "bad" parts of a city as where "bad" people live is precisely to misunderstand what might contribute to the occurrence of antisocial acts, how they might be prevented or minimised in the future, and how discrimination can skew an understanding of criminal actions and actors. A neat, humanistic attribution of moral blame to the unified, knowing, and autonomous individual, in short, risks recapitulating the logic that justified

the Dickensian slum and which bought-in to social Darwinism as an explanation of the social standing of, for example, poor, queer, and non-white
citizens. To try to understand the huge variety of uncontrolled factors which
might contribute to individuals' actions, adding together currently siloed
disciplinary enquiries, would instead be a move toward better understanding human experience and how future harms might be reduced. But it would
also mark a huge shift in a deeply ingrained sense of justice; it is maybe the
biggest ask that posthumanism could make, and yet one that it is inherently
unlikely to be able to escape making.

The biological, internal, and environmental factors discussed above show
how humans can be shaped by the things which not only surround them,
but pass through them and become a part of them. We might also add further concerns from empirical psychology and neuroscience,[11] building from
Libet et al.'s (1983) study of movement which suggested that the intent to
act can come *after* an action has been initiated, troubling the rational, knowing subject as the originator of action. Soon et al. more recently

> found that the outcome of a decision can be encoded in brain activity
> of prefrontal and parietal cortex up to 10 [seconds] before it enters
> awareness. This delay presumably reflects the operation of a network
> of high-level control areas that begin to prepare an upcoming decision
> long before it enters awareness.
>
> (Soon et al. 2008, 543)

Derk Pereboom and Greg Caruso, in their discussion of punishment in the
wake of the critique of moral responsibility, also note that

> work in psychology and social psychology on automaticity, situation-
> ism (see Doris) and the adaptive unconscious . . . reveal just how wide
> open our internal psychological processes are to the influence of exter-
> nal stimuli and events in our immediate environment.
>
> (Pereboom and Caruso 2018, 7)

Similarly, the humanities, where most explicitly posthumanistic work is
currently being undertaken, have a long history of engagement with wider
social factors such as unconscious and discursive influences and biases.[12]
From the legacies of Freudian and Lacanian psychoanalysis to the diag-
nosing of the textual and social effects of all forms of discrimination and
inequality, humanities scholars have sought to show how development and
discourse impact upon human cultures. But moral responsibility still
dominates; to diagnose how psychology and society distort our perceptions
and experiences is vital work, but the full implications, as part of a wider

understanding of influences on human action, are missed if these insights are only used to ascribe blame. As Bennett puts it,

> sometimes moral outrage . . . is indispensable to a democratic and just politics. . . . Outrage will not, and should not, disappear, but a politics devoted too exclusively to moral condemnation and not enough to a cultivated discernment of the web of agentic capacities can do little good.
>
> (Bennett 2010, 38)

There are so many things about which we should be outraged. The mainstreaming of far-right propaganda; the calculated denial of climate crises; the extent of police brutality; the spikes in hate crime in the wake of the election of Donald Trump, the referendum on Great Britain's exit from the European union, and the outbreak of Covid-19 (see Edwards and Rushin; Devine; Gover et al.) – what would it require to understand the actions of the individuals involved? Not to relate to them, but to know that you're susceptible to similar (if, luckily, better) influences? Waller often tries to face up to this:

> We are a profoundly hierarchical species . . . [and frequently] this tendency is manifested in a way that is brutal and horrific. . . . [W]hen asked, most people vehemently deny that they would go along with the authoritative orders to shock others [as in the Milgram conformity experiments[13]], though experimental studies show that most of those who sincerely make that denial are honestly but profoundly mistaken. When there is a genuine fault that afflicts most of us and that we certainly did not choose or create and of which most of us are blissfully unaware, it is difficult to suppose that we deserve blame for that fault, but it is not at all difficult to conclude that it is a very severe fault that has – from the Crusades to Abu Ghraib – produced terrible wrongs.
>
> (Waller 2011, 162)

The validation of our worst assumptions/the support of our best intentions; external pressures to conform/the encouragement to explore; the negative/positive shaping of attitudes and perceptions through the cultures we inhabit, these are all (un)lucky factors that can contribute to our actions. But I feel, keenly, the sense of injustice that might come in asking you to reconsider the responsibility of the perpetrators of criminal acts (and it doesn't feel balanced by knowing that this argument applies as much to undeserved praise as to blame). How useful an ally is a posthumanism which reopens the question of responsibility, even if it doesn't doubt the badness of the acts; even if it would only want to understand in order to work for change for the better?

But when it comes to the rider in the hospital, awaiting blame and punishment, there are unavoidable things to consider. A culture that fetishises speed; a bike that can vastly exceed the speed limit; a work day which raised anxiety and stress; the need to be home to take an important call; poor depth perception leading to misjudgement; the bad luck of character that Waller describes; unconscious choices already made in the seconds and fractions of seconds before he was even aware of his decision; a head full of toxo; a destructive microbiome; a history of lead toxicity and poor nutrition linked to an impoverished upbringing – these and innumerable other unknown factors could have contributed to his action. It was a bad thing to do, and real harm was done, but blame feels more difficult, and moral responsibility more challenging to attribute.

Conclusion

In attempting to understand what contributes to our actions, contemporary sociological, criminological, biological, economic, environmental, neuroscientific, and psychological hypotheses already provide a model for thinking through a posthumanistic approach to querying moral responsibility; they are already exploring the things which are both unexpected constituents of us and out of our control. And as we start to add together all of these avenues of research, the exceptions to the unified, knowing, and autonomous human actor might quickly become the rule; the coherent, rational, and free subject starts to seem like a myth, and all of these weird parasites, poisonings, and influences seem far less extraordinary.

In his definition of "posthumanism" for the *Posthuman Glossary*, Cary Wolfe states that

> not only is the line between human and non-human impossible to definitively draw with regard to the binding together of neurophysiology, cognitive states and symbolic behaviours, the line between "inside" and "outside," "brain" and "mind," is also impossible to draw definitively. . . . In a fundamental sense . . . what makes us "us" is precisely not us; it is not even "human." . . . What we call "we" is in fact a multiplicity of relations between "us" and "not us," "inside" and "outside."
> (Wolfe 2018, 358)

I agree, but what typically goes unspoken in these sorts of assessment is that they mean the troubling question of moral responsibility has already forced the door. In that realisation, however, there can be a call to action: a demand to recognise the gamut of harms and their effects, and to plan for amelioration, but one which exists alongside a radical compassion and a rethinking of justice and our rationales for punishment. It's a shocking ask, but one already asked.

Notes

1. This passage continues:

> yet we do not really believe that there is such a thing [as this core self]. . . . It works, but it does not exist. This puts the modem defender of our moral and political order in something of a difficult position.

Hill offers an overview of this discussion in law, though outside of a posthumanist framework.

2. Even if this is the normative model, this is a more complex question in legal theory, and particularly with regard to the justness of retributive punishment; see, for example, Raz's *From Normativity to Responsibility* for more on responsibility, and Alec Walen's excellent overview of "Retributive Justice." A humanistic stance, however, remains central to many legal models, see, for example, "Morissette v. United States":

> The contention that an injury can amount to a crime only when inflicted by intention is . . . as universal and persistent in mature systems of law as belief in freedom of the human will. . . . Crime . . . constituted only from concurrence of an evil-meaning mind with an evil-doing hand, was congenial to an intense individualism and took deep and early root in American soil.

For further discussion see also Berman, "Punishment and Justification."

3. For example: "Unscrupulous prosecutors have long realized that if the case against a defendant is weak, then grisly photos of the crime scene and the murder victim . . . can be an effective substitute for the missing evidence: when jurors are outraged by a criminal act, their desire to strike back at somebody can easily overwhelm their concern about whether the defendant is the appropriate target of their wrath" (Waller, 11).

4. For foundational works see, for example, Williams, *Moral Luck* and Nagel, *Moral Questions*.

5. Sometimes termed "constitutive luck." There is also "present luck," for example,

> an agent's mood . . . situational features of the environment, how aware she is of the morally significant features of her surroundings, . . . whether . . . attention wanders at just the right/wrong moment[, etc.]. . . . The one-two punch of constitutive luck . . . and present luck completely undermine basic-desert moral responsibility.
>
> (Caruso and Dennett)

6. "If we don't play, we are banished . . . from the human community. Given that alternative, it is hardly surprising when even those who are most severely mistreated . . . insist that they want to remain in the game" (Clark, Dennett, and Waller). See also Caruso to Dennett:

> Studies show, for instance, that low socioeconomic status in childhood can affect everything from brain development to life expectancy, education,

incarceration rates, and income. The same is true for educational inequity, exposure to violence, and nutritional disparities. It's a mistake . . . to think that luck averages out in the long run – it does not.

(Caruso and Dennett)

7. Influential ideas in this vein for contemporary posthumanism include Bruno Latour's "actor network theory" (e.g. 2005), Karen Barad's "intra-action" (2007), Jane Bennett's "vibrant matter," and the work of, for example, Donna Haraway (e.g. 2016), Anna Tsing (e.g. 2015), Rosi Braidotti (e.g. 2013), and Cary Wolfe (e.g. 2010).

8. Haraway, it's worth noting, rejects "posthumanism" as a term "too easily appropriated" by teleological tranhumanists embracing "technoenhancement" (Gane, 140); she does also acknowledge, however, that "[l]ots of people doing posthumanist thinking . . . don't do it that way" (140).

9. See, for example, Nessa Carey's *The Epigenetics Revolution* for an introduction to the field.

10. "Crime" can be hard to define and study; an area's arrest rates, rather than being an indicator of elevated criminal behaviour, can instead be read as a marker of institutional racism, over-policing, and/or historic bias. See, for example, Yessenia Funes, "Air Pollution May Affect Crime, But the Problem Goes Deeper Than That."

11. For example, Zeki et al. "For the Law, Neuroscience Changes Nothing and Everything" or the recently inaugurated Neuroscience and Philosophy of Free Will project (https://neurophil-freewill.org).

12. For an overview of empirical research on social influence see Cialdini and Goldstein's "Social Influence."

13. See, for example, Milgram's "Group Pressure and Action Against a Person." For a more recent discussion of the legacy of Milgram's work, see Haslam and Reicher's "50 Years of 'Obedience to Authority.'"

Bibliography

Barad, Karen. 2007. *Meeting the Universe Halfway*. Durham NC: Duke University Press.

Belsey, Catherine. 2014. *The Subject of Tragedy: Identity and Difference in Renaissance Drama*. Abingdon: Routledge.

Bennett, Jane. 2010. *Vibrant Matter*. Durham, NC: Duke University Press.

Berman, Mitchell N. 2008. "Punishment and Justification." *Ethics* 118(2): 258–90.

Braidotti, Rosi. 2013. *The Posthuman*. Cambridge: Polity Press.

———. 2019. *Posthuman Knowledge*. Cambridge: Polity Press.

Cameron, Lindsay P., Angela Nazarian, and David E. Olson. 2020. "Psychedelic Microdosing: Prevalence and Subjective Effects." *Journal of Psychoactive Drugs*: 1–10.

Carey, Nessa. 2011. *The Epigenetics Revolution*. London: Icon Books.

Carranza, Nancy. 2018. "Agency." *Geneaology of the Posthuman*, April. https://criticalposthumanism.net/agency/.

Cialdini, Robert B., and Noah J. Goldstein. 2004. "Social Influence: Compliance and Conformity." *Annual Review of Psychology* 55(1): 591–621.

Clark, Tom. 2012. "Singling Out the Agent." *Naturalism.org*, April. www.naturalism. org/resources/book-reviews/singling-out-the-agent.

Clark, Tom, Daniel Dennett, and Bruce Waller. 2012. "Exchange on Waller's *Against Moral Responsibility.*" *Naturalism.org*, October. www.naturalism.org/resources/ book-reviews/exchange-on-wallers-against-moral-responsibility.

Cook, Thomas B., et al. 2015. " 'Latent' Infection With *Toxoplasma gondii*: Association With Trait Aggression and Impulsivity in Healthy Adults." *Journal of Psychiatric Research* 60: 87–94.

Cryan, John, and Siobhan M. O'Mahony. 2011. "The Microbiome-Gut-Brain Axis: From Bowel to Behavior." *Neurogastroenterology & Motility* 23: 187–92.

Dennett, Daniel. 2012. "Dennett Review of *Against Moral Responsibility.*" *Naturalism.org*, October. www.naturalism.org/resources/book-reviews/dennett-review-of-against-moral-responsibility.

Dennett, Daniel, and Gregg Caruso. 2020. "On Free Will: Daniel Dennett and Gregg Caruso Go Head to Head." *Aeon*, March 14, 2020. aeon.co/essays/ on-free-will-daniel-dennett-and-gregg-caruso-go-head-to-head.

Devine, Daniel, 2018. "The UK Referendum on Membership of the European Union as a Trigger Event for Hate Crimes." February 5, 2018. http://dx.doi.org/10.2139/ ssrn.3118190.

Dinan, Timothy, Roman Stilling, Catherine Stanton, and John Cryan. 2015. "Collective Unconscious: How Gut Microbes Shape Human Behavior." *Journal of Psychiatric Research* 63: 1–9.

Doris, John M. 2002. *Lack of Character: Personality and Moral Behavior*. Cambridge: Cambridge University Press.

Edwards, Griffin Sims, and Stephen Rushin. 2018. "The Effect of President Trump's Election on Hate Crimes." January 14, 2018. http://dx.doi.org/10.2139/ ssrn.3102652.

Feigenbaum, James J., and Christopher Muller. 2016. "Lead Exposure and Violent Crime in the Early Twentieth Century." *Explorations in Economic History* 62: 51–86.

Ferrando, Francesca. 2019. *Philosophical Posthumanism*. London: Bloomsbury.

Flegr, Jaroslav. 2013. "How and Why Toxoplasma Makes Us Crazy." *Trends in Parasitology* 29(4): 156–63.

Flegr, Jaroslav, et al. 2002. "Increased Risk of Traffic Accidents in Subjects With Latent Toxoplasmosis: A Retrospective Case-Control Study." *BMC Infectious Diseases* 2(1): 11.

Funes, Yessenia. 2019. "Air Pollution May Affect Crime, But the Problem Goes Deeper Than That." *Gizmodo*, May 12. https://earther.gizmodo.com/ air-pollution-may-affect-crime-but-the-problem-goes-de-1839781306.

Gane, Nicholas. 2006. "When We Have Never Been Human, What Is to Be Done? Interview with Donna Haraway." *Theory, Culture & Society* 23(7–8): 135–58.

Gilbert, Scott F., Jan Sapp, and Alfred I. Tauber. 2012. "A Symbiotic View of Life: We Have Never Been Individuals." *The Quarterly Review of Biology* 87(4): 325–41.

Gover, A. R., et al. 2020. "Anti-Asian Hate Crime During the COVID-19 Pandemic: Exploring the Reproduction of Inequality." *American Journal of Criminal Justice* 45: 647–67.

Haraway, Donna. 2016. *Staying With the Trouble*. Durham: Duke University Press.

Haslam, Alexander S., and Stephen D. Reicher. 2017. "50 Years of 'Obedience to Authority': From Blind Conformity to Engaged Followership." *Annual Review of Law and Social Science* 13(1): 59–78.

Hill, John Lawrence. 1997. "Law and the Concept of the Core Self: Toward a Reconciliation of Naturalism and Humanism." *Marquette Law Review* 80(2): 289–390.

House, Patrick K., et al. 2011. "Predator Cat Odors Activate Sexual Arousal Pathways in Brains of *Toxoplasma gondii* Infected Rats." *PLoS One* 6(8).

Johnson, Stefanie K., et al. 2018. "Risky Business: Linking *Toxoplasma Gondii* Infection and Entrepreneurship Behaviours Across Individuals and Countries." *Proceedings of the Royal Society B: Biological Sciences* 285(1883).

Latour, Bruno. 2005. *Reassembling the Social*. Oxford: Oxford University Press.

Libet, Benjamin., Curtis A. Gleason, Elwood W. Wright, and Dennis K. Pearl. 1983. "Time of Conscious Intention to Act in Relation to Onset of Cerebral Activity (Readiness-Potential): The Unconscious Initiation of a Freely Voluntary Act." *Brain* 106(3): 623–42.

Ling, Vinita J., et al. 2011. "*Toxoplasma Gondii* Seropositivity and Suicide Rates in Women." *The Journal of Nervous and Mental Disease* 199(7): 440–44.

Lochner, Lance, and Enrico Moretti. 2004. "The Effect of Education on Crime: Evidence from Prison Inmates, Arrests, and Self-Reports." *American Economic Review* 94(1): 155–89.

Lu, Jackson G., et al. 2018. "Polluted Morality: Air Pollution Predicts Criminal Activity and Unethical Behavior." *Psychological Science* 29(3): 340–55.

Mielke, Howard W., and Sammy Zahran. 2012. "The Urban Rise and Fall of Air Lead (Pb) and the Latent Surge and Retreat of Societal Violence." *Environment International* 43: 48–55.

Milgram, Stanley. 1964. "Group Pressure and Action Against a Person." *The Journal of Abnormal and Social Psychology* 69(2): 137–43.

Muller, Christopher, Robert J. Sampson, and Alix S. Winter. 2018. "Environmental Inequality: The Social Causes and Consequences of Lead Exposure." *Winter Annual Review of Sociology* 44(1): 263–82.

Nagel, Thomas. 1979. *Moral Questions*. New York: Cambridge University Press.

"Neuroscience and Philosophy of Free Will." *Neuro Philosophy of Free Will*, 2020. neurophil-freewill.org.

O'Dell, Bonne Ford, Emily V. Fischer, and Jeffrey R. Pierce. 2019. "The Effect of Pollution on Crime: Evidence From Data on Particulate Matter and Ozone." *Journal of Environmental Economics and Management* 98.

Pereboom, Derk, and Gregg Caruso. 2018. "Hard-Incompatibilist Existentialism: Neuroscience, Punishment, and Meaning in Life." In *Neuroexistentialism: Meaning, Morals, and Purpose in the Age of Neuroscience*, edited by Gregg D. Caruso and Owen Flanagan. http://dx.doi.org/10.2139/ssrn.2758312.

Ramsbotham, Lord David, and Bernard Gesch. 2009. "Crime and Nourishment: Cause for a Rethink?" *Prison Service Journal* 182: 3–9.

Raz, Joseph. 2014. *From Normativity to Responsibility*. Oxford: Oxford University Press.

Rees, T., Bosch Thomas, and Douglas Angela. 2018. "How the Microbiome Challenges our Concept of Self." *PLoS Biology* 16(2).

Reyes, Jessica Wolpaw. 2015. "Lead Exposure and Behaviour: Effects on Antisocial and Risky Behaviour Among Children and Adolescents." *Economic Inquiry* 53: 1580–605.

Samojłowicz, Dorota, Aleksandra Borowska-Solonynko, and Marcin Kruczyk. 2017. "New, Previously Unreported Correlations Between Latent *Toxoplasma gondii* Infection and Excessive Ethanol Consumption." *Forensic Science International* 280: 49–54.

Sampson, Robert J., and Alix S. Winter. 2018. "Poisoned Development: Assessing Childhood Lead Exposure as a Cause of Crime in a Birth Cohort Followed Through Adolescence." *Criminology* 56: 269–301.

Sapmaz, Şermin Yalın, et al. 2019. "Relationship Between *Toxoplasma gondii* Seropositivity and Depression in Children and Adolescents." *Psychiatry Research* 278(2019): 263–67.

Sapolsky, Robert. 2009. "TOXO: A Conversation with Robert Sapolsky." *Edge.org*. www.edge.org/conversation/robert_sapolsky-toxo.

Soon, Chun Siong, Marcel Brass, Hans-Jochen Heinze, and John-Dylan Haynes. 2008. "Unconscious Determinants of Free Decisions in the Human Brain." *Nature Neuroscience* 11: 543–45.

Supreme Court of the United States. 1952. "Morissette v. United States." *United States Reports* 342.

"Toxoplasmosis." *NHS Choices*. NHS, 2017. www.nhs.uk/conditions/toxoplasmosis.

Tsing, Anna. 2015. *The Mushroom at the End of the World*. Princeton: Princeton University Press.

Valles-Colomer, Mireia, et al. 2019. "The Neuroactive Potential of the Human Gut Microbiota in Quality of Life and Depression." *Nature Microbiology* 4(4): 623–32.

Walen, Alec. 2020. "Retributive Justice." In *The Stanford Encyclopedia of Philosophy (Fall 2020 Edition)*, edited by Edward N. Zalta. https://plato.stanford.edu/archives/fall2020/entries/justice-retributive/.

Waller, Bruce. 2011. *Against Moral Responsibility*. Cambridge, MA: MIT Press.

Webster, Colin, and Sarah Kingston. 2014. *Poverty and Crime*. London: Joseph Rowntree Foundation.

Williams, Bernard. 1981. *Moral Luck: Philosophical Papers 1973–1980*. Cambridge: Cambridge University Press.

Wolfe, Cary. 2018. "Posthumanism." In *Posthuman Glossary*, edited by Rosi Braidotti and Maria Hlavajova, 356–58. London: Bloomsbury, 2018.

———. 2010. *What Is Posthumanism?* Minneapolis: University of Minnesota Press.

Wright, John Paul, et al. 2008. "Association of Prenatal and Childhood Blood Lead Concentrations with Criminal Arrests in Early Adulthood." *PLoS Medicine* 5(5).

Xiao, Jianchun, et al. 2018. "*Toxoplasma gondii*: Biological Parameters of the Connection to Schizophrenia." *Schizophrenia Bulletin* 44(5): 983–92.

Zeki, Semir, et al. 2004. "For the Law, Neuroscience Changes Nothing and Everything." *Philosophical Transactions of the Royal Society of London. Series B: Biological Sciences* 359(1451): 1775–85.

9 The Filter Problem for Posthuman Bioethics

The Case of Hyperagency

David Roden

Posthumanism can be critical or speculative in orientation. Both forms are critical of human-centred (anthropocentric) thinking. However, their rejection of anthropocentricism applies to different areas: critical posthumanism (CP) rejects the anthropocentrism of modern philosophy and intellectual life; speculative posthumanism (SP) opposes human-centric thinking about the long-run implications of modern technology.

While critical posthumanists are interested in the posthuman as a cultural and political condition, speculative posthumanists are interested in a possibility of certain technologically created nonhuman agents.

Speculative posthumanism (SP) can be spelt out in a single sentence:

> (SP) *Wide* descendants of current humans could *cease to be human* by virtue of a history of technical alteration.

I've coined the term "wide descent" because an exclusive consideration of "narrow" biological descendants as candidates for posthumanity would be excessively restrictive given our moral concern with the implications of technically enhanced "transhuman" successors or various forms of synthetic and artificial life (Roden 2014, 208, 209–13). For example, in the case of a recursive intelligence explosion speculated upon by Vernor Vinge and others, the resultant posthumans would be synthetic agents whose creation might owe nothing to ordinary processes of biological reproduction. Consequently, the process of *ceasing to be human* should not be understood in exclusively biological terms – as ceasing to belong to a particular species – since not all candidates for posthuman status would be biological individuals, but might include artificial or engineered entities.

According to a formula I call the "Disconnection Thesis" *ceasing to be human* is better understood as ceasing to belong to the system of technical uses arranged around the needs of humans (Roden 2012, 2014, 105–23). Disconnection would be a process by which technical things become

DOI: 10.4324/9781003020707-13

"feral," leaving our human system of ends to carve out independent careers. I have argued that the Disconnection Thesis provides us with a conceptual handle on human–posthuman difference without telling us what posthumans would be like. This is as it should be. *For posthumans, there are none.* The DT is a *mechanism-independent description* that leaves the field of posthuman natures to be determined empirically. For example, a revolution in machine intelligence could conceivably engineer a disconnection. But even supposing machine superintelligence is not possible, a disconnection might be generated by an entirely different mechanism!

Accelerated Ethics

SP and DT extract the "moral core" of the singularity hypothesis and similar futurist scenarios. What matters about disconnection is not *just* the mechanism by which it comes about but the conceptual possibility that something made for our system of ends might withdraw from that system and discover ends autonomously.

This process might not work out so badly. For example, in a classic 1981 paper, the philosopher Paul Churchland imagines people whose brains are linked by a microwave channel that mediates inter-brain communication as effectively as the corpus callosum integrates parts of the same brain. "Once the channel is opened between two or more people" he writes:

> they can learn (learn) to exchange information with the same intimacy and virtuosity displayed by your own cerebral hemispheres. Think what this might do for hockey teams, and ballet companies, and research teams!
>
> (Churchland 1981, 88)

A neural interface that allowed brains to communicate with one another as well as the two halves of our cerebral hemispheres might lead to a "neuroculture" that frees itself from the "stupid, limiting spoken language" that so frustrates cylon Cavil in *Battlestar Galactica*. A neuroculture might be much less violent than human culture, since pain, along with agency, might be distributed among participating bodies (Roden 2014, 3). A world in which human brains are suffused in microwaved harmony might, thus, be a *Neurotopia* with particular excellences in team sports and performing arts.

If the neuroculture were limited to a small group of volunteers it might provide an effective disconnection mechanism, excluding its members from any human society that might still be around. After all, the cognitive, affective, and communicative capacities of such linked brains might be radically different to those of unlinked humans.

Again, it is possible that such an interface would never work – or at least not in the way that Churchland envisages. But this would not undermine SP if some other mechanism turned out to be feasible. Perhaps some part of humanity might become genetically modified to be super-cooperators, or super-individualistic predators (think artificial vampires!). So, again, technologically constituted agents might break out of the human system for reasons other than those actuating the AI breakout described in Vinge's singularity scenario.

The take home moral of this is (once more) that the truth of SP is not dependent on the feasibility of any one technology. Because disconnections might be realised in multiple ways, it is a weaker claim than the Singularity Hypothesis and thus has greater plausibility. Nonetheless, it is a routine commonplace of apocalyptic SF that disconnection might not always work out!

This possibility is vividly and comically portrayed in Charles Stross' 2006 novel *Accelerando*. *Accelerando* begins in a twenty-first-century Europe buzzing with speculative technologies and madcap utopianisms, but its timeline extends into a dark post-singularity future dominated by self-improving artificial intelligences. These are wide human descendants of human corporations and automated legal systems which have achieved sentience and legal personhood. As one character ruefully observes, the phrase "smart money" has taken on an entirely new meaning!

Eventually, these "corporate carnivores" – known by the epithet "Vile Offspring" – institute a new form of capitalism known as "Economics 2.0" in which supply and demand relationships are computed too rapidly for those burdened by a narrative chain of personal consciousness to keep up. E 2.0 is so remorselessly efficient that it comes to dominate a major part of the solar system. Whole planets are pulverised and diverted to fast-thinking clouds of smart matter "blooming" around the sun (Stross 2006, 208–10). Stross' post-singularity scenario certainly seems bad for humans. Even their souped-up transhuman cousins prove incapable of functioning within E 2.0 and can only flee to the outer solar system and beyond as the Earth and its neighbours are "ethnically cleansed."

Stross' fiction goes well beyond facile apocalypticism, however, for it raises some nice ethical questions about the moral status of posthumans. Some moralists might deny that E 2.0 is really "good" for posthumans in a way that could compensate for its bad impact on humans. Assume (with Stross) that Vile Offspring lack personal consciousness. Many would infer from this that they are not the kinds of entities whose lives can go well or badly. They are not worthy of moral consideration at all. Consequently, there would be no gain in Vile Bliss to compensate for the human misery caused by the obliteration of planets. However, I think this evaluation of nonpersons is unwarranted when, as in the context of posthuman bioethics, *the totality of possible minds or agents* is under consideration. To support this claim I'll set out one among several arguments which imply that our grip on subjectivity or rationality *cannot impose robust epistemic filters on posthuman (or alien) forms of agency or subjectivity*. On this basis I will argue that (ante-Disconnection) we cannot exclude the possibility that there could be non-personal life forms like the Vile Offspring that are *no less morally considerable than persons*.

We can understand this epistemic limitation in terms of the distinction between an anthropologically bounded posthumanism and an anthropologically *unbounded* posthumanism (ABP and AUP). ABP holds:

1. There are unique conditions of possibility C for agency that any posthuman successor to humans must satisfy (C are transcendental constraints)
2. Agents satisfying C can know that they are agents and can deduce a priori that they satisfy C (C are reflectively accessible to agents satisfying C)
3. Humans typically satisfy C

ABP's import becomes clearer if we consider the collection of histories whereby posthuman wide descendants of humans could feasibly emerge. I refer to this set as Posthuman Possibility Space (PPS – see Roden 2014, 53). Given that posthumans would be agents of *some* kind (see Chapter 6) and given ABP, members of PPS would have to satisfy the same transcendental conditions (C) on agency as humans.

Daryl Wennemann assumes something along these lines in his book *Posthuman Personhood*. He adopts the Kantian idea that agency consists in the capacity to justify one's actions according to reasons and shared norms. For Wennemann, a person is a being able to "reflect on himself and his world from the perspective of a being sharing in a certain community" (Punzo 1969, cited in Wennemann 2013, 47). This is, for him, a condition of posthuman agency as much as of human agency. It implies that, whatever the future throws up, posthuman agents will be social and, arguably, linguistic beings, even if they are robots or computers, have strange bodies, or even stranger habits. If so, PPS *cannot* contain non-anthropomorphic entities whose agency is significantly nonhuman in nature. Consequently ABP implies that there are a priori limits on posthuman *weirdness*.

AUP, by contrast, leaves the nature of posthuman agency to be settled *empirically* (or technologically) and thus we are deceived in thinking that we have an a priori understanding of the space of agents and minds. Here, an illustration will suffice: It can be argued there are certain rational norms that generalise from actual human persons to hypothetical posthumans given that the latter would be intelligent, goal-driven systems. Such beings would thus count as rational subjects able to evaluate their behaviour according to explicit goals and generalisable rules (Brassier 2011). Consider this plausible practical maxim:

RD: If a rational subject x has some categorical value v^*, x should want to be motivated by the desire that v^* is realised.

This maxim is just designed to make explicit that it would be irrational for an agent who (say) had as their overriding value "making paperclips" not to want to be motivated by *that* desire rather than desires whose realisation would not maximise paperclips, production. We would say of an agent who asserted that making paperclips is an overriding value for them but was indifferent about whether they would be motivated by incompatible desires in the future that they just did not understand the concept of value.

Steve Omohundro has argued on the basis of this principle that we can predict the overriding goals of even post-singularity entities. For example, systems with an unlimited capacity to alter their software or physical structure would have an incentive to make modifications that would help them achieve their goals more effectively. Such being would qualify, in Michael Sandel's terminology, as "hyperagents" – for whom "every constitutive aspect of agency (beliefs, desires, moods, dispositions and so forth) is subject to . . . control and manipulation" (Danaher 2014, 227; Sandel 2004). In what follows, I'll use Sandel's term as a portmanteau to refer to any being (biological or otherwise) which exhibits this extreme form of plasticity.

According to Omohundro, robotic hyperagents would want to ensure that such alterations do not threaten their current goals:

> So how can it ensure that future self-modifications will accomplish its current objectives? For one thing, it has to make those objectives clear to itself. If its objectives are only implicit in the structure of a complex circuit or program, then future modifications are unlikely to preserve them. Systems will therefore be motivated to reflect on their goals and to make them explicit.
>
> (Omohundro 2008)

I think this picture of values that are self-present in the physical component of a self-modifying agent is questionable in a way that is salutary for bioethicists (such as Wennemann, Omohundro and Nick Bostrom) who think human norms of practical reasoning can be extended beyond a posthuman-making event. Although not all hyperagents need to be purely artificial (they might be radically enhanced humans or synthetic life forms, etc.), the practical reasoning assumptions are not robust enough to generalise from artificial to other kinds of agents, as we shall see.

Omohundro requires that there could be internal system states of post-singularity AIs whose value content could be legible for the system's internal probes. Obviously, this assumes that the properties of a piece of hardware or software can determine the content of the system states that it orchestrates independently of the external environment in which the system is located.[1] Let's assume it does. The problem for Omohundro is that the relevant inner states are liable to depend on their relations to other inner states. The intrinsic shape or colour of an icon representing a station on a metro map is arbitrary. There is nothing about a circle or a square or the colour blue that signifies "station." It is only the conformity between the relations between the icons and the stations in metro system it represents which does this (Churchland's 2012 account of the meaning of prototype vectors in neural networks utilises this analogy – see section 4.1). Thus, the value meant by an internal state s under some configuration of the system must depend on some inner context (like a cortical map) where s is related to lots of other states of a similar kind (Fodor and Lepore 1992).

But relationships between states of the self-modifying AI systems are assumed to be extremely plastic because each system will have an excellent model of its own hardware and software and the power to modify them (I refer to this attribute as "hyperplasticity" in Roden 2014, 100–2.) If these relationships are modifiable *then any state could exist in innumerable alternative configurations*. It might function like homonyms within or between languages, having very different value-meanings in different contexts.

Suppose that a hyperplastic AI wishes to ensure a state in one of its value circuits, s, retains the value content it has under the machine's current configuration: v^*. To do this it must avoid altering itself in ways that would lead to s being in an inner context in which it meant some other value (v^{**}) or no value at all. It must clamp itself to those contexts to avoid s assuming v^{**} or v^{***}, etc. To achieve clamping, though, the AI needs to ensure its self-modifying activity only produces configurations of itself in which s is paired with a context that preserves its meaning.

The problem for the AI is that all $[s + c]$ pairings are just more internal system states, which, themselves, might assume different meanings in different contexts. To ensure that s means v^* in context c it needs to have done to some $[s + c]$ what it had been attempting with s – restrict itself to the supplementary contexts in which $[s + c]$ leads to s having v^* as a value and not something else.

Now, a hyperagent will always be in a position to modify any configuration that it finds itself in (for good or ill). So this problem will be replicated for any combination of states $[s + c \ldots + \ldots]$ that the machine could assume within its configuration space. Each of these states will have to be repeatable in yet other contexts, etc. Clamping any arbitrary s so that it keeps on meaning value v^* requires that we have already clamped some undefined set of contexts for s, and this condition applies to that set, and *its* possible contexts, in turn. Since each combination of system states is a system state to which the principle of contextual variability applies recursively, *there is no final system state for which this problem does not arise recursively.*

So, when Omohundro envisages a machine scanning its internal states to explicate their values he seems to be proposing that an infinite task has already been completed by a being with vast but presumably finite computational resources. If this argument is sound, we have good reason to think that a hyperagent would never have a reliable method with which to ensure that its agency conforms to a given categorical value or desire. This is because none of its internal states could be relied upon to "clamp" that desire in place in its future configurations. Could such a system have second order desires of the kind specified in maxim **RD**? Well, if the system were highly rational it would "know" that it could not ensure that any desire could survive some self-modification.

Ironically, the hyperagential elimination of common-sense psychological categories follows from arguments that, historically, were designed to secure it against encroachments from physicalism or materialism: in other words, forms of moderate anti-reductive materialism such as Donald Davidson's anomalous monism or Daniel Dennett's stance account of intentionality. Desires might supervene on internal states but given the plausible context sensitivity of content, they would not show up in probes of *non-intentionally characterised system states* (Davidson 2001a, 222). If this moderate anti-reductionism applies generally (as it seems it must, given the holism of content determining features discussed earlier) then the argument generalises straightforwardly from artificial wide-descendants such as Omohundro's self-modifying robots, to all hyperagents, including perhaps our enhanced biological successors.

So, a rational hyperagent would differ fundamentally from the rational subjects we think we know: the maxim **RD** would not be applicable for it. More generally and radically, attributing desires to other hyperagents would not be a helpful way of predicting their actions because desire attributions would not be robust under future iterations of the same hyperagent – some self-modifying tweak could always delete or alter its current desires, or engender new ones.

How about beliefs? Well, here the situation seems even worse. On most conceptions of rational subjectivity, epistemically rational believers should adjust their beliefs only in the light of good reasons. For example, I believed that the conference where I originally read this essay as a paper was taking place on Lesbos while I was giving it. There are conceivable (if exotic) circumstances that might have led me to subsequently abandon this belief. But (assuming that I was rational) they would have had to involve me acquiring some well-supported countervailing belief: for example, the experience of waking up in an immersive virtual reality tank in an industrial park on the outskirts of Birmingham (or the Matrix).

But if beliefs locally supervene on internal structure without there being any psychophysical theory that reduces belief types to non-intentionally characterised structural types, no hyperplastic could ensure its doxastic commitments would survive some future self-tweak of non-intentionally characterised internal states. So (once more), it seems that attributing beliefs, desires, wishes, hopes other intentional states could not be a pragmatically effective means for a hyperplastic agent to understand either itself or other hyperagents.

The Parochialism of Bioethics

The upshot of this is that the framework of belief-desire psychology and the reasoning practices that facilitate our interpersonal life would be largely inapplicable to beings with significant hyperagency. They would not be situated in the "space of reasons" because their flexibility would far outrun its predictive powers. Note that this argument does not stem from the assumption that the practical reasoning supported by belief-desire psychology is only applicable to humans or could not be exercised by nonhumans. For example, we can situate non-hyperplastic (merely plastic) nonhuman animals, nonhuman persons, or even autonomous devices within the space of reasons by adopting the "intentional stance" towards them, as Dennett proposes (1995). However, the intentional stance has no efficacy where an agent's powers have been extended to the threshold of hyperagency (wherever or however that may occur).

So, either we must conclude that the very idea of "hyperagent" is oxymoronic or that we have encountered a kind of singularity in the space of reasons – a place where merely human folk conceptions of agency and rationality no longer apply. The first position may be arguable, but it has the counter-intuitive consequence that arbitrarily extending the power of an agent to manipulate its world must, at some point, expunge agency rather than produce a different state of agency.

This follows from a strong anthropocentric and anti-realist position regarding agency: to wit, that anything human subjects cannot interpret as an agent isn't one.[2] However, the second (anthropocentrism) is vulnerable to the degree that the first is weak. In what follows, then, I will assume a posthumanist (and realist) position that countenances agents that would be uninterpretable for ordinary humans and could be produced by incrementing agential prowess to the threshold of hyperplasticity.

I think this posthumanist/realist position on agency comports nicely with Scott Bakker's proposal that our post-Kantian framework for thinking about agency does not guarantee insight into the structural properties of real agents or those of hypothetical ones. It is a developed version of a highly specific set of heuristic tools, evolved to deal, as he says, with a very specific "problem ecology":

> Our socio-cognitive systems are the ancient product of particular social environments, ways to optimize our biomechanical interactions with our fellows in the absence of any real biomechanical information. Our ancestors also relied on them to understand their macroscopic environments, to theorize nature, and it proved to be a misapplication. Nature in general is not among the things that social cognition can solve (though social cognition can utilize nature to solve social problems, as seems to the case with myth and religion). Only ignorance of nature qua natural allowed us to assume otherwise.
>
> (Bakker 2014, see also 2017)

Not only does intentional thinking not extend to non-mental nature, its writ may not run beyond the conditions obtaining in our current, fast-changing socio-technological niche. To wantonly project it in bioethical debate is to fall into a transcendental illusion that mistakes "a subjective necessity of a connection of our concepts . . . for an objective necessity in the determination of things in themselves" (Kant 1978, A297/B354). Here, the illusion consists in the extension of putatively transcendental conditions beyond the niche where they have any predictive or interpretative efficacy.

If we identify persons as creatures to whom maxims of practical reason of the kind discussed earlier are efficacious, it follows that posthumans who qualify as hyperagents might not be subject to them and, in such cases, could not qualify as either persons or beings that combine the attributes of personhood with moral powers giving them superior moral status to persons ("supra-persons"). They would be agents that we are not in a position to understand currently. Thus, while we cannot justifiably ascribe them "supra-personal" moral status, neither would it be epistemically rational to assume their moral status is intrinsically inferior to that of persons (Buchanan 2009).

A priori reflection framed in folk psychological terms can consequently issue in no substantive ethical norms for such posthumans. From this, we should conclude that the human bioethical discussion about – say – whether hyperagency would enhance or destroy the meaning or "giftedness" of life is subject to this very anthropocentric error and is liable to be inapplicable in regions of PPS inhabited by the hypothetical agents whose possibility raises such moral concerns.

Both supporters of enhancement like Danaher or critics like Michael Hauskeller have consequently overestimated the relevance of human enhancement ethics to posthumans because they have failed to consider whether their shared conceptual framework for thinking about agency could be rendered irrelevant by technological alterations that might induce a hyperagential disconnection (Hauskeller 2011). This is why speculative posthumanism, although ethically relevant – insofar as it allows us to "think" the inhuman horizon of current human technical practice – must recuse itself from any direct ethical responsibility (see Roden 2019, 2020).[3]

If non-personal posthumans are possible, it follows that a disconnection (a posthuman maker) might erase the "manifest image" in which persons encounter each other in a common, intersubjective world. Following this "hyperapocalypse" (Bakker's equivalent term, coined in his novel *Neuropath*, is "semantic apocalypse" – see Bakker 2010), there could be modes of life that we could not understand without becoming similarly posthuman (assuming, here, that any agent possesses an at least pre-reflexive grasp of its mode of existence).

Given this assumption, only hyperagents would be qualified to evaluate their ethical situation while *our* bioethical understanding of hyperagential virtue and bliss would have to await our own disconnection from the human system of ends through acquisition of hyperagency. That is, the bioethics of posthumanity remains in a condition of transcendental illusion until we are in a better epistemic situation than is currently the case: when we are in a position to encounter or, better, *become* the relevant kind of posthuman. In the case of posthuman ethics, engineering necessarily pre-empts a priori philosophising.

Allowing for this severe epistemic restriction, it remains legitimate to speculate that hyperagential entities like Stross' Vile Offspring may have morally considerable states (Vile Bliss) that would be intrinsically valuable in ways that we cannot comprehend within folk-psychological categories such as pleasure, pain, ecstasy, and gratitude; or modes of "agency" that are (in some absolute sense) equivalent but utterly alien to the rational agency of Kantian moral subjects.

Their lack of personal consciousness would imply only that comparing Vile impersonal states in terms of human personal states would be ethically unsupportable. Having a phenomenology, as I have argued elsewhere, does not entail a correlative understanding of what a phenomenology *is* or the ability to generalise it within the space of possible agents or minds (Roden 2013).

Some might regard this concession as morally objectionable – indicative of a bad character even. Surely, only a terminally sick animal could forgo judgement regarding its biological and semantic extinction! However, I do not take the prospect of us becoming hyperagents or some other radically disconnected kind of posthuman lightly; I just claim that we don't how to reliably evaluate it (yet).

Conclusion

If some threshold of hyperagency is possible, then a condition of its possibility may be the practical elimination of the folk-psychological mode of self-understanding (belief-desire psychology) that is necessary for hyperagency to be a topic of philosophical concern for human ethicists. Such speculation can issue no prescriptions since – absent actual hyperagents – our moral discourse is articulated around principles that might not apply to hyperagents at all. To overcome this restriction, bioethicists need to come clean and explicitly impose filters on posthuman possibility. Such a priori constraints may be supported by assuming some variant of anthropologically bounded posthumanism. However, the hypothetical conditions attributed to exotic forms of posthumanity, like hyperagents, place ABP in doubt and thus warrant the adoption of unbounded posthumanism (see Roden 2019, 2020).

Notes

1. This property of non-environmental determination is known as "local supervenience" in the philosophy of mind literature. If local supervenience for value-content fails, any inner state could signify different values in different environments. Clamping machine states to current values would then entail restricting the situations in which a particular mind might find itself.
2. This anti-realist construal of agency concepts is enshrined in Davidson's observability assumption: namely that an observer can under favourable circumstances tell what beliefs, desires, and intentions an agent has (Davidson 2001b, 99).
3. For a detailed discussion of this the issue in respect of the analytic pragmatism of Wilfred Sellars and Robert Brandom, see Roden 2017.

Bibliography

Bakker, R. Scott. 2010. *Neuropath*. New York: Tor.
———. 2014. Accessed September 25, 2014. http://rsbakker.wordpress.com/2014/09/08/arguing-no-one-wolfendale-and-the-penury-of-pragmatic-functionalism/.
———. 2017. "On Alien Philosophy." *Journal of Consciousness Studies* 24(1–2): 31–52.

Brassier, Ray. 2011. "The View From Nowhere." *Identities: Journal for Politics, Gender and Culture* 17: 7–23.

Buchanan, Allen. 2009. "Moral Status and Human Enhancement" *Philosophy & Public Affairs*, 37: 346–81.

Churchland, Paul M. 1981. "Eliminative Materialism and the Propositional Attitudes." *The Journal of Philosophy* 78(2): 67–90.

Churchland, Paul. 2012. *Plato's Camera: How the Physical Brain Captures a Landscape of Abstract Universals.* Cambridge, MA: MIT Press.

Danaher, John. 2014. "Hyperagency and the Good Life – Does Extreme Enhancement Threaten Meaning?" *Neuroethics* 7(2): 227–42.

Davidson, Donald. 2001a. *Essays on Actions and Events*, Vol. 1. Oxford: Oxford University Press.

———. 2001b. *Subjective, Intersubjective, Objective*, Vol. 3. Oxford: Oxford University Press.

Dennett, Daniel C. 1995. "Do Animals Have Beliefs?" *Comparative Approaches to Cognitive Science* 111.

Fodor, Jerry, and Ernest LePore. 1992. *Holism: A Shopper's Guide.* Oxford: Blackwell.

Hauskeller, Michael. 2011. "Human Enhancement and the Giftedness of Life." *Philosophical Papers* 40(1): 55–79.

Hayles, N. Katherine. 1999. *How We Became Posthuman: Virtual Bodies in Cybernetics, Literature, and Informatics.* Chicago, IL: University of Chicago Press.

Kant, Immanuel. 1978. *Critique of Pure Reason.* Translated by Norman Kemp Smith. New York: St Martin's Press. (Originally published in German as *Kritik der reinen Vernunft* [Johann Hartknoch, 1787]).

Omohundro, Stephen M. 2008. "The Basic AI Drives." In *Proceedings of the First AGI Conference, 171, Frontiers in Artificial Intelligence and Applications*, edited by Pei Wang, Ben Goertzel, and Stan Franklin, 483–92. Amsterdam: IOS Press.

Roden, David. 2012. "The Disconnection Thesis." In *The Singularity Hypothesis: A Scientific and Philosophical Assessment*, edited by Ammon H. Eden, Johnny H. Søraker, James H. Moor and Eric Steinhart, 281–98. London: Springer.

———. 2013. "Nature's Dark Domain: An Argument for a Naturalised Phenomenology." *Royal Institute of Philosophy Supplements* 72: 169–88.

———. 2014. *Posthuman Life: Philosophy at the Edge of the Human.* London: Routledge.

———. 2017. "On Reason and Spectral Machines: Robert Brandom and Bounded Posthumanism." In *Philosophy After Nature*, edited by Rosi Braidotti Rick Dolphijn, 99–119. London and New York: Rowman & Littlefield International.

———. 2019. "Subtractive-Catastrophic Xenophilia." *Identities: Journal for Politics, Gender and Culture* 16(1–2): 40–46.

———. 2020. "Posthumanism: Critical, Speculative, Biomomorphic." In *The Bloomsbury Handbook of Posthumanism*, edited by Jacob Wamburg and Mads Thomsen, 81–94. London: Bloomsbury.

Sandel, Michael J. 2004. "The Case Against Perfection." *The Atlantic Monthly*, Inc, Boston.

Stross, Charles. 2006. *Accelerando*. New York: Ace.
Vinge, Vernor. 1993. "The Coming Technological Singularity." Accessed June 2011. http://www-rohan.sdsu.edu/faculty/vinge/misc/WER2.html.
Wennemann, Daryl J. 2013. *Posthuman Personhood*. New York: University Press of America.
Williams, Bernard. 1973. "The Makropulos Case: Reflections on the Tedium of Immortality." In *Problems of the Self: Philosophical Papers*, 1956–72. Cambridge: Cambridge University Press (this edition 1992).

Index